Evolution and Revolution:
American Society, 1600–1820

Evolution and Revolution

AMERICAN SOCIETY, 1600–1820

James A. Henretta
University of Maryland, College Park

Gregory H. Nobles
Georgia Institute of Technology

D. C. HEATH AND COMPANY LEXINGTON, MASSACHUSETTS / TORONTO

Cover: "Old State House, Boston," by J. B. Marston,
Massachusetts Historical Society.

Published simultaneously in Canada.

Printed in the United States of America.

International Standard Book Number: 0-669-08304-6

Library of Congress Catalog Card Number: 86-81264

To Patricia Wilson and Rebecca and Emily Henretta

To Phoebe and Sarah Nobles

Preface

This book is both a second edition of James Henretta's *The Evolution of American Society* (published by D.C. Heath in 1973) and a new work of historical interpretation. It employs the interdisciplinary approach of the first edition, yet it incorporates the impressive results of the historical scholarship of the past decade. A second new aspect of this book is that it has a co-author, Gregory Nobles, who has taken primary responsibility for one-half of the text. *Evolution and Revolution* blends his perspective and the data and interpretations advanced by other scholars into an original theoretical and conceptual structure. The result is a new synthesis of the early American experience.

Evolution and Revolution is intended for an undergraduate audience. A generation of graduate students profited from the pioneering synthesis of the "new social history" presented in the first edition. Many historians outside the field of early American history enjoyed its broad sweep and engaging generalizations. A fair number of teachers liked it well enough to assign it in their undergraduate classes. Some undergraduates understood the approach and the arguments advanced in *The Evolution of American Society,* but many others found them too difficult. They could not easily comprehend chapters that were organized conceptually rather than chronologically; often they lacked sufficient historical knowledge to follow the logic of complex interrelationships and interpretations.

Our revision addresses these classroom deficiencies of the first edition. *Evolution and Revolution* has ten short chapters rather than six long ones. The chapters are organized in a chronological as well as a topical fashion. Moreover, a strong political narrative enhances the presentation in many chapters, providing students with a coherent "story" of American development. In addition, we have often focused on specific individuals, families, or groups to illustrate—and enliven—our analysis of the process of social and economic change.

The human dimension of the American experience has received more emphasis because of recent scholarly trends. Influenced by the discipline of anthropology, historians have produced outstanding studies of the cultures of native Americans, women, enslaved blacks, Chesapeake planters, and New England farmers. By including this new material and perspective, we have expanded the original interdisciplinary approach and added a new cultural dimension to the historical narrative. We hope that *Evolution and Revolution* will be as intellectually stimulating to the students of the 1990s as the first edition was to their teachers during the 1970s.

Christine Heyrman and John Brooke read the book in manuscript and helped us to improve it in many ways. We wish to thank them and the scores of historians whose scholarship has informed our understanding of the history of early America.

James A. Henretta

Gregory H. Nobles

Preface to THE EVOLUTION OF AMERICAN SOCIETY (1973)

Most books about American history deal primarily with *public* events, with the activities of individuals as they are manifest in the formal institutions of the economy, society, or polity. If they are included at all, historical data relating to family life, religious and cultural activities, or social life are relegated to a few separate and prosaic chapters. Thus segregated and subordinated, this rich and vital material lies inert and lifeless on the page; it does not exert leverage on the crucial questions of historical interpretation.

This dichotomy between private experience and public activity is artificial and misleading. The lives of men and women constitute an integrated and indivisible whole, as does the multidimensional social space in which they work out the possibilities of their existence. In all societies, both past and present, the personal and private is also the public and political.

This fundamental insight has long been obscured by the paucity of theoretical or conceptual models that would permit the historian to visualize the social system as a single interrelated whole. Only those writers who benefited from the analytical technique bequeathed to posterity by Karl Marx were able, in some measure, to escape this fragmented (and therefore distorted) view of historical reality.

With the utilization by growing numbers of historians of the methods and assumptions of the modern social sciences, the situation has been dramatically altered. Demographic analyses of past populations have begun to reveal basic patterns of birth, marriage, and death and to hint at the psychological and cultural implications of particular forms of family life and household organization. At the same time, anthropological studies of cultural units and sociological investigations of class, status, and power have revealed the intimate connection that exists between the family and the wider structure of community life. Other studies of political organization, of psychological maturation, and of economic change have disclosed the myriad threads of causation that link the various facets of social existence. In the end, the underlying unity of all aspects of human life has emerged with startling clarity.

With this conceptual advance, the elucidation of pattern, configuration, and direction has replaced the study of discrete acts, facts, and events as the major concern of historical and social analysis. It is no longer sufficient simply to describe what happened, or even to "explain" individual occurrences in terms of cause and effect. This approach is too limited, too static, too linear. The new vision of reality has created a new method and a new vocabulary; we speak of the political *process,* of economic *development,* of the *context* of social action, of ecological *relationships,* of the *direction* of social change. The emphasis is one of motion, not inertia; and the assumption is that the lives of individual historical actors can only be understood in terms of a larger dynamic whole. Thus the structure and the content of this book: the evolution (the growth and structural differentiation) of an entire social system through time.

Contents

PLATES

MAPS

CHARTS

Evolution and Revolution:
American Society, 1600–1820

An Overview

> The poor man's conscience is clear; yet he is ashamed. . . . He feels himself out of the sight of others, groping in the dark. Mankind takes no notice of him. He rambles and wanders unheeded. In the midst of a crowd, at church, in the market . . . he is in as much obscurity as he would be in a garret or a cellar. He is not disapproved, censured, or reproached; *he is only not seen*.
>
> John Adams, *Diary*

> To study the laws of history we must completely change the subject of our observation, must leave aside kings, ministers, and generals, and study the common infinitesimally small elements by which the masses are moved. No one can say in how far it is possible for man to advance in this way toward an understanding of the laws of history; but it is evident that only along that path does the possibility of discovering the laws of history lie.
>
> Leo Tolstoy, *War and Peace*

History is the study of men and women struggling to work out their individual lives within a larger process of change. Some change is cyclical, reflecting elemental regularities in nature. Day changes into night, winter into spring, one year into the next. In the past, these natural rhythms determined the pattern of human life. Mankind slept when natural light failed, planted and reaped as the seasons dictated, and struggled against the rigors of the environment to maintain life itself. These basic and demanding regularities no longer constitute the framework of American life. Millions of Americans still measure the passage of time and events by the seasonal clock of agricultural life, but these people are now a distinct minority. Their habits, customs, and way of life do not reflect the experience of the nation as a whole.

This was not always the case. Between 1600 and 1820, the overwhelming majority of Americans — white and black, poor and rich — lived on the land. Their lives were very different from ours; yet their accomplishments and their failures are of importance to us. Settlers from Europe bought or seized the American land from its aboriginal Indian occupants. These colonists created new societies whose character and ideals still influence our lives. Some of these men and women fought to establish the United States as an independent republic. The work of many others initiated the long process that transformed the new nation into an urban industrial society.

In 1820, the era of industrial development had barely begun. The United States boasted few large cities and even fewer large factories. Yet great changes had already occurred in American life. The population had increased from a few thousand in 1620 to 1 million in 1740. Then the number of Americans doubled each generation, reaching 2.5 million in 1775, 5.3 million by 1800, and 9.6 million by 1820. The alteration in the economic system was equally impressive. In less than two centuries, American plantations and farms became the primary suppliers of wheat, tobacco, and cotton to European markets. And American merchants managed one of the largest fleets of ships in the entire world.

Social change in America was not only cyclical and repetitive but also cumulative and irreversible. Small gains in population, in economic development, and in political maturity were compounded from decade to decade and from generation to generation. By 1740, American society was much larger and more diverse than it had been a century before; and it was further changed by 1820.

The organization of this book reflects that historical reality. It stresses those events and processes that prompted these patterns of cumulative change. Between 1600 and 1740, English settlers slowly built a new society. They formed authoritarian and deferential political and religious institutions. They bore large numbers of children, accepted substantial numbers of German, Scotch-Irish and other non-English migrants, saw their ranks grow through the migration of tens of thousands of indentured servants, and imported equally large numbers of African slaves. Eventually this population growth, economic expansion, and cultural diversity undermined the stability of their institutional order.

Between 1740 and 1775, the old order was first challenged and then overthrown. Initially, religious dissenters questioned traditional beliefs and systems of authority. Soon political agitators — backcountry farmers, Scotch-Irish migrants, and eventually Patriot merchants and artisans — attacked existing governmental institutions and practices. By the time of the Declaration of Independence, the values and institutions that had evolved over the first century and a half of American history lay in disarray.

Society had to be created anew by an act of political will. For the next generation, the character of the new republican order was discussed and debated wherever men and women gathered — in private houses and public taverns, in town meetings and legislative assemblies. Idealist revolutionaries deemed many traditional practices, such as a legally privileged aristocracy, to be incompatible with republican values. They wrote new constitutions and devised new political institutions. On the whole, these innovations enhanced the political power of the people in relation to government officials and guaranteed the religious and legal rights of individual citizens. Public institutions and values conformed more closely to the ways in which Americans wanted to act (and were acting) than did the older stratified and deferential political and social system. The American Revolution created a new equilibrium between private action and public ideology. Between 1776 and 1820 a new order came into being.

Revolutionary ideology was a source of tension as well. In a nation formally dedicated to the principles of liberty and equality, hundreds of thousands of blacks were still held in slavery, American women lacked full legal equality, and the society's wealth continued to be distributed unequally. In some respects, this disparity between the myths of the new republic and its actual social and economic structure had grown more acute by 1820. An emergent capitalistic economic system raised the American standard of living and created a middle-class culture. At the same time, this new mode of economic production and social relations created its own inequalities in wealth, status, and power. The American Revolution had run its course; the Industrial Revolution had begun.

But what of the actors in this cumulative historical drama? Thousands, even millions, of people appeared on the stage of American life between 1600 and 1820. Each man and woman had a unique personality; yet many of them shared cultural norms and social values. As a result, they often acted in similar ways, what Tolstoy called "the common infinitesimally small elements" that make up the patterns of historical experience. For example, seventeenth-century New England parents had seven or eight children. Three generations later, most of their descendants limited the number of their offspring to three or four. Such decisions, which are made freely and individually yet with similar and uniform timing, have continuously built up the broad structures of American life.

The actions of thousands of "unheeded" and obscure men and women, as John Adams called them, shaped America's historical destiny in the most obvious ways as well. They built a prosperous economy, transformed American cultural values by embracing evangelical religious doctrines, and fought a long and bitter war to establish independent republican polities. Their deeds established the context of political and social life. To understand these ordinary lives is to comprehend the limits within which kings and governors and patriots and presidents have acted. In all societies, past and present, the lives of the great and those of the humble are inextricably intertwined.

PART I

The Formation of American Society, 1600–1740

1

The Collision of English and Native American Cultures

The United States began as a set of colonies, established at a particular moment in English history. Seen from that perspective, the history of America between 1600 and 1650 is essentially the story of transplanted English men and women. Some of them fled to New England to escape religious persecution; others migrated to Virginia or Maryland in the hope of financial gain or as a result of dire economic necessity. Their lives in America were fundamentally shaped by the Puritan religious ideology or the adventurous economic attitudes that they brought with them.

Seen from another perspective, the history of seventeenth-century America is one of invasion and conquest. Aboriginal peoples had populated the continent of North America for centuries and had evolved a host of small but stable tribal societies. Suddenly, white-skinned Europeans — English, French, Dutch, and Spanish — invaded their lands, primarily along the eastern and southern seacoasts. The confrontation — military, economic, and biological — nearly destroyed the culture of many tribes.

This conflict of cultures shaped many aspects of life in the new English colonies. Indian attacks threatened the very existence of the first settlements and deeply affected their political and social development. For more than a generation, English settlers huddled in small fortified towns and followed the dictates of authoritarian leaders. At the same time, the invaders used Indian crops, technology, and furs to ensure their economic survival. From the very outset, the fates of Europeans and native Americans were deeply intertwined.

Taken together, the three themes of economic and religious transplantation, racial confrontation, and cultural interaction formed the contours of American history in the early seventeenth century.

England in 1600

In 1600, England was in the midst of a major change that extended throughout the society. At the apex of power, Queen Elizabeth I (1558–1603) neared the end of a long and eventful reign. The highly venerated Virgin Queen (so-called because she never married) had rallied her Protestant subjects to repulse a Spanish Catholic invasion in 1588. Aided by fierce storms sweeping off the North Atlantic, her swift naval forces smashed the Armada of King Philip II of Spain. This crucial victory prevented the restoration of the authority of the Roman Catholic Pope over the Church of England.

Threats to the established church came from within England as well. When King Henry VIII established a national Protestant church, he retained many Catholic rituals and institutions. Elizabeth I did likewise. As head of the church, she defended the elaborate rituals of many of its services and supported the authority of the religious hierarchy of bishops. She ignored the calls of many ministers for a representative (or presbyterian) system of church government. Presbyterian clergy and laity demanded the elimination of bishops and the creation of elected synods. The queen also repudiated radical religious sects such as the Puritans. Puritans attacked both the hierarchical structure and the elaborate ceremonies of the Church of England. They placed religious authority in the hands of the lay congregation and followed plain forms of worship.

Religious controversy broke out with renewed fervor under Elizabeth's successor, James I (1603–1625). The new king was the anonymous author of *The True Laws of Free Monarchy* (1598). This tract argued that God had given kings a "divine right" to rule over the affairs of church as well as state. Shortly after ascending to the throne, James put his ideas into practice. He condemned Puritanism, demanding that all of his subjects conform to the doctrines and rituals of the Church of England. In addition, James sought to increase the political power of the monarchy. He declared his intention to impose taxes and laws without the consent of the Houses of Parliament, the representative political bodies dominated by landed aristocrats and country gentlemen. To create this absolute monarchy, James courted the support of the merchant community.

Since 1500, English merchants had grown increasingly wealthy and powerful. They began by selling high-quality English wool to European textile manufacturers. They next organized an elaborate putting-out system of cloth production. The putting-out system utilized merchant capital and the labor of rural families. Merchant wagons collected wool from sheep farmers and delivered it to thousands of poor families. These families lived in small cots (cottages), grew their food in tiny gardens, and worked long hours for low wages. The women and children cleaned the wool and spun it into yarn, which male weavers wove into cloth on household looms. The merchants sold some of this cloth in urban areas and the rest in foreign markets, creating a favorable balance of trade.

Elizabeth I greatly expanded the English textile trade by giving substantial financial bounties and special privileges to these merchant capitalists. She granted the Levant Company a royal charter that gave it a monopoly of trade with Turkey in 1581 and awarded similar privileges to the Guinea (or Africa) Company in 1588 and to the East India Company in 1600. The queen also supported the privateering attacks of Sir Francis Drake against ships carrying gold and silver to Spain from Mexico and Peru. And she granted charters to Sir Walter Raleigh and other aristocratic adventurers to establish privateering bases along the southeastern coast of North America.

The English queen likewise looked favorably on foreign expansion. Since the 1570s, English pamphlet writers such as Richard Hakluyt (pronounced Hack-lit) had been urging state support for the colonization of North America. He urged Protestant clergy to save the "savages" both from their pagan beliefs and from conversion to Spanish Catholicism. And he tempted merchants with the prospect of lucrative trade with the native peoples. Aristocrats were the first to heed Hakluyt's call. They hoped to establish new feudal estates in America, to grow rich and powerful on the labor of subject Indian peoples and English tenant farmers. Between 1570 and 1600, Elizabeth I granted land in America to various titled aristocrats, but they lacked the financial resources to sustain colonies in far-distant America. Sir Fernando Gorges failed to sustain an English outpost in Maine, and colonists dispatched by Sir Walter Raleigh disappeared without a trace in the famous "lost" colony of Roanoke, North Carolina.

James I also sought an American empire, but he placed his hopes on the merchant community. In 1606, the king granted charters to trading companies based in Plymouth and London to settle "that part of America, commonly called Virginia." The merchants who founded the Virginia Company of London hoped to find gold, to engage in profitable trade with the Indians, or to raise crops to sell in European markets. Their plans melded self-interest with the national interest. By establishing colonies in America, the English would avoid dependence on other nations for raw materials and would have more goods to exchange for Spanish gold. For his part, the king was to receive one-fifth of all precious metals found in Virginia and sovereignty over a vast new dominion.

The king's quest for power also colored his domestic political policies. A strong-willed, even arrogant man, King James styled himself as one of the new breed of absolute monarchs that ruled many of the nations of Europe. Like other absolutist kings, such as Philip II of Spain, James attacked the power of the aristocratic nobility. These nobles were the descendants of medieval feudal lords. They owned vast estates and had special legal privileges, such as their own political body — the House of Lords, the upper house of Parliament. By creating new aristocratic titles for his own supporters, James sought to undermine the status and power of the traditional nobility.

Many of the old noble families were already declining for economic reasons. Prices in England (and throughout Europe) had nearly doubled between 1550 and 1650. The influx of gold and silver from the mines of Spanish America was one cause of this inflation. After centuries of stability, the money supply of Europe nearly doubled; prices rose as people used this extra money to bid up the value of fixed amounts of land, food, and manufactured goods. Rapid population growth was another, and perhaps the primary, cause of the rise in prices. Following the Black Death of 1347, the European population stagnated for two centuries and then increased rapidly. The number of English men and women rose from 3 to 5 million between 1550 and 1620, vastly increasing the demand for scarce resources.

The inflation in prices temporarily undermined the financial position of many aristocrats. Most nobles rented out their estates to peasant tenants on long leases, often of fifty years' duration. As prices shot up, aristocrats trapped in these long-term contracts continued to receive the same amounts as rent. Since their income remained stable while many of their expenses doubled, their real income was cut in half.

Meanwhile, the inflation benefited other landowning classes. Many members of the gentry increased their wealth. The gentry was composed of prominent gentlemen who owned small estates and represented their localities in the House of Commons, the lower house of Parliament. Most gentry families rented out their land on short-term leases and so could raise their rents as prices rose. Many yeomen farmers prospered as well. These farmers and their families owned and worked their own land, sometimes with the help of a hired servant. They amassed capital by selling wheat, rye, and meat at ever increasing prices. Some yeomen used their new wealth to acquire more land or to expand their houses and barns. What one historian has called a Great Rebuilding transformed the domestic architecture of rural England. Finally, some tenant farmers were able to improve their social position. These peasants paid the contractual rents to their aristocratic landlords and used their surplus funds to buy land and become freeholders.

These economic changes created a state of flux at the top of the traditional rural social hierarchy. Local aristocrats still dominated the affairs of hundreds of English villages, but their power was no longer absolute. Gentry and yeomen families resisted aristocratic power from below, while James I and his merchant allies challenged it from above.

The situation at the bottom of the agricultural pyramid was equally chaotic. Around 1450, English landowners began to raise more sheep in order to meet the growing demands of wool merchants. To create sheep pastures, the landowners terminated the leases of tenant farmers. More important, they used legal means to break up common-field peasant communities. Over the centuries, the peasants had won rights of customary use over this land and the privilege of managing most village affairs through communal institutions. Using their political influence in Parliament, landowners secured legislation that ended peasant use-rights and allowed the enclosure of this arable land with fences or hedges for sheep pasturing. In 1500, common-field peasant communities occupied 70 percent of the arable land of England; by 1650, they controlled only 50 percent.

The Enclosure Movement pushed tens of thousands of peasants off their ancestral lands. Some displaced peasants worked as wage laborers on large estates. Others became the landless cotters who spun and wove wool into cloth. Still other peasants moved about the English countryside with their families in search of work. Many young men and women ended up in London, which grew spectacularly from 75,000 inhabitants in 1550 to 200,000 in 1600 and 325,000 in 1650. Thousands of other peasants arrived in the western seaports of Plymouth and Bristol. Poor and without prospects, they eagerly signed labor contracts (called indentures) offered by merchants and sea captains. These contracts usually required a servant to work for a master for five years. In return, the servant received food, clothing, lodging, and — most important of all — free passage to the new English colonies in America. Between 1600 and 1700, at least 100,000 young English men and women sought a better life in the American settlements as indentured servants. After their indentures expired, they hoped to become yeomen farmers — a status beyond their reach in England.

Thus, the peculiar allocation of economic power in England in 1600 determined the institutional structure of the new American colonies. The aristocracy was in relative decline. The failed ventures of Gorges and Raleigh indicated that the aristocracy lacked the financial or psychological resources to impose its social vision on the first English settlements in the New World. On the other hand, the English monarchy and its merchant allies in London and Plymouth were active and aggressive. As a result, English expansion took the form of trading ventures and led to the creation of colonies that exported the agricultural staples of sugar and tobacco.

Religious and social divisions within England also affected the composition of American settlements. Religious sects that dissented from the doctrines of the Church of England sought refuge in America. Puritans settled in New England, Catholics in Maryland, and Quakers in Pennsylvania. Moreover, population growth and the Enclosure Movement created a mass of impoverished peasants in England. "Our country [is] overspread," a magistrate in the county of Kent complained, "not only with unpunished swarms of idle rogues and counterfeit soldiers but also with numbers of poor and weak but unpitied servitors." Alarmed by these hordes of "masterless" men and women, politically powerful landlords supported the colonization movement. As a result, the new American colonies would attract hundreds of thousands of settlers (and not merely a few hundred missionaries or merchants). The legacy of religious diversity and social disruption in England thus shaped the first century of American history.

From Corporation to Polity in Virginia

The first permanent English settlers in North America arrived as members of joint-stock companies. London merchants had used these business organizations for decades in trading with European and Mediterranean countries. As members of a joint-stock company, merchants agreed to abide by certain rules and procedures. They pooled contributions of money, ships, goods, and men and settled on how to divide the profits of the venture among the investors. The Virginia Company of London organized the first settlement of Jamestown, Virginia, in 1607 along these lines. Under the terms of the royal charter, the investor-stockholders controlled the company. Some stockholders contributed money (paying the substantial sum of £12 for each share of stock); others received shares by settling in the colony. Each share of stock entitled the holder to 100 acres of the company's land and, after seven years, a share of the trading profits.

The stockholders were optimistic. For years, English seafarers had brought back glowing reports of the southern coast of North America. For example, in 1585, Arthur Barlowe wrote enticingly of the islands in Roanoke Sound:

> The earth bringeth forth all things in abundance, as in the first Creation, without toil or labor. . . . [The islands are] most beautiful and pleasant to behold, replenished with deer . . . hares and divers beasts, and about them the goodliest and best fish in the world and in greatest abundance.

This image of easy abundance in an American Eden lured investors to the company, and the notion that these riches could be gathered "without toil or labor" led adventurers to expect a life of luxury.

Hopeful expectations faded away in the face of dreadful experiences. By 1619, the Virginia Company had failed as a business enterprise. Most of the shareholder-adventurers who went to Virginia had died, as had most of the 2500 servants dispatched by the company to assist them. Disease took the lives of scores of colonists on overcrowded merchant ships; food shortages and diseases in Virginia swept away most of those who had survived the voyage.

"There were never Englishmen left in a foreign country as we were in this new-discovered Virginia," wrote one adventurer, "the pitiful murmuring and outcries of our sick men [continued] every night and day. In the morning their bodies were trailed out of the cabins like dogs to be burned." "So the truth is," another migrant concluded, "we live in the fearfullest age that ever Christians lived in."

The environment itself posed the most deadly threat to the colonists. In moving from Europe to America, the settlers became exposed to new bacteria and viruses against which their bodies had no natural resistance or immunities. Moreover, the particular sites of the first English settlements — coastal estuaries at the mouths of rivers — were especially dangerous. The mixture of salt and fresh water in standing pools and marshes became stagnant and filthy during the summer months, making an excellent breeding ground for harmful bacteria and mosquitoes. Frequently, the salt level in the water was as high as five times what it should be in healthful drinking water. Weakened by a long ocean voyage and a deficient diet, the company's settlers fell prey to a variety of afflictions, ranging from salt poisoning to mosquito-borne malaria to epidemics of fly-borne dysentery and typhoid fever. Only a few settlers survived long enough to become "seasoned" by developing resistance to these hazards. And even these lucky ones did not live very long. Most 20-year-old men or women who came to the Chesapeake Bay area died well before turning 50; by comparison, residents of England (and of New England) usually lived twenty years longer.

To attract new migrants to the hard-pressed colony, the Virginia Company changed its land policy. Beginning in 1618, settlers received private title to 50 acres of land for each adult family member or servant transported to the colony. In addition, the company bestowed huge land grants on influential Englishmen to induce them to invest and settle in Virginia. To make Virginia even more attractive to property owners, the company created a representative assembly in Jamestown in 1619. Twenty-two elected members sat in the House of Burgesses. Together with Governor George Yeardley and his council of advisers, who were appointed by the company, the burgesses passed laws regulating local affairs.

The formation of the House of Burgesses established the principle of partial self-rule on the part of settlers in North America. When James I dissolved the Virginia Company in 1624, following charges of mismanagement, and made Virginia into a royal colony, he allowed this representative body to continue. His brother, Charles II (1625–1642), adopted a similar policy when he issued a charter for the new colony of Maryland in 1632. Charles directed Lord Baltimore, the Catholic proprietor, to make laws "with the advice, assent, and approbation of the Free-Men."

These policies, along with the successful cultivation of tobacco, transformed Virginia from a company-run disaster into a ruthless private enterprise. There was an almost insatiable demand for tobacco in England, where people smoked it for its pleasurable, slightly narcotic effect. Most Virginia tobacco was of a lower quality than that produced in the Spanish Caribbean, but quality hardly seemed to matter. It seemed that English consumers would take whatever they could get from their Virginia connection. In 1616, the planters of the Chesapeake Bay region sent a mere 2500 pounds of tobacco to England, but the limited supplies only drove up the price. By 1640, planters in Virginia and Maryland were exporting 3 million pounds of tobacco each year.

Tobacco gave the Chesapeake region what both English merchants and local settlers wanted — a New World crop that was prized but not produced in England. So promising were the prospects of tobacco production that the inhabitants of Virginia organized their lands and even their lives around it. Rather than clustering together in small villages for safety or sociability, they isolated themselves on individual plantations along the waterways, in order to ensure easy transportation for their tobacco. Company officials and later observers criticized this "unhappy Form of Settlement" that scattered the population "without any Rule or Order." However, given the impetus of the tobacco boom, the planters arranged themselves on the landscape in a way that made economic sense.

Virtually everyone in the Chesapeake region grew tobacco, and the "vile weed" (as James I called it) became the main medium of exchange. As one observer of Virginia's multiplying plantations put it in 1623, "nothing is done in any one of them, but all is vanished into smoke." Tobacco required constant care and tending but produced a good yield. Virginia planters used their crop to cover debts with each other and to buy needed goods — tools, food, liquor, and clothes — from English and Dutch ship captains. Rising production repeatedly drove tobacco prices down, but Virginia remained a boom colony; the planters, quite literally, grew their own money. Working alone, some men were able to grow as much as 1000 pounds of tobacco a year.

If a man could take advantage of the labor of other men, he could produce even more. In fact, the control of labor shaped the early social development of Virginia and Maryland. High tobacco profits encouraged planters to expand production by adding workers; however, high mortality continually depleted the available labor force. In order to increase — or merely replenish — the labor force, planters relied on a steady stream of white indentured servants. As tobacco became the Chesapeake region's main export, human beings became its main import. Power flowed into the hands of those who — by means of wealth from England or the exercise of skill in America — acquired large land grants and the servants to work on them. Ralph Wormeley, an influential settler, had gained title to more than 3000 acres of land by the time of his death in 1650. Three years later, his widow Agatha, remarried to Sir Henry Chicheley, had seventeen servants at work on her estate.

The lives of indentured servants were hard, much harder than those of hired laborers in England. Determined to grow wealthy, planters worked their servants to death or sold their contracts to the highest bidders. "My master Atkins," wrote Thomas Best, a young English migrant who had signed a five-year labor contract to get to Virginia, "hath sold me for 150 sterling pounds like a damned slave." John Rolfe, the adventurer who promoted the extensive planting of tobacco, noted widespread complaints against "the Governors, Captaines, and Officers in *Virginia*: for buying and selling men and boys. . . ." In 1620, the Virginia Company itself shipped some single servant women to the colony, selling each for 120 pounds of high-quality tobacco. While professing concern for the "libertie of marriage," the company directed that these women be married "only to such freemen or tenants as have means to maintain them." In fact, these women — like their male counterparts — were just commodities to be sold on the labor market. Captain John Smith thought it was "odious" that masters sold male and female servants "for forty, fifty, or three-score pounds, whom the Company hath sent over for eight or ten pounds at most, without regard to how they shall be maintained."

As the planters acquired land and labor, they sought to rule the colony according to their own interests. Their quest for power quickly produced conflicts with royal officials and migrant gentlemen. Perhaps as many as one-third of the first adventurers in Jamestown were gentlemen, and many of the colony's first settlers had been drawn from the highest ranks of English society. George Perry, son of the Earl of Northumberland, ventured to the new colony, as did George Sandys, son of the Archbishop of York. In 1620, eight members of the governor's council in Virginia had attended the great universities of Oxford and Cambridge or received legal training at the Inns of Court in London. Such men felt they were born to rule; they demanded deference not only from impoverished servants and yeomen but also from those well-to-do planters who had risen up from more humble origins.

However, by 1635, "new men" dominated the council and forced Sir John Harvey to resign the governorship. Some of the new councillors brought wealth and high status with them from England. Thus, Samuel Mathews had twenty-three servants working on his plantation within a few years after he arrived in the colony. However, the majority of councillors rose up from middling ranks. They transported only one or two indentured servants to Virginia, but then prospered through hard work and the aggressive pursuit of their self-interest. When Harvey arrested one ambitious councillor, John Utie, on "suspicion of Treason to his Majestie," Utie refused to accept his authority. A confident, self-made man, Utie responded with a firm "and wee the like to you sir" and forced Harvey's return to England.

This episode did not reflect the existence of political equality or liberty in Virginia. Harvey's ouster instead confirmed the dominant authority of a local oligarchy of servant-owning and tobacco-planting families. An influx of new men of wealth and status augmented the authority of this evolving elite. Between 1645 and 1665, ambitious younger sons of London merchants and English country gentlemen migrated to Virginia. Many had inherited shares of stock in the Virginia Company. Thomas Culpeper, who migrated in 1649, had a father, uncle, and cousin among the members of the original company. And William Byrd, who established the famous dynasty of political leaders, inherited extensive lands from his mother's family.

Because of the activities of Governor William Berkeley, this budding aristocracy was soon in firm control of the political institutions of Virginia. Berkeley arrived in the colony in 1642. Exploiting family investments in Virginia, political connections in England, and his success as an Indian fighter, Berkeley served as governor from 1642 to 1652 and then again from 1660 to 1681. To solidify his authority, Berkeley awarded land grants and lucrative government offices to influential members of the Virginia oligarchy. By the 1660s, Berkeley's "Green Spring" faction (so named after the governor's country estate) dominated the council and ruled the society in a manner that served its own interests.

As in England itself, the reality of rule by the elite lay behind a facade of representative institutions in Virginia. But, in other respects, the new colony was very different from England and had even evolved away from its own origins. The joint-stock company had given way to a settler-society functioning under a royal governor. Self-made men who had ruthlessly exploited the labor of indentured servants held political power, alongside of aggressive descendants of the original shareholders of the Virginia Company. As the colonists sought to fulfill "the rumors of plenty to bee found at all time[s] in Virginia," they created a uniquely exploitative economic and political order.

The Puritan Commonwealth in New England

As the Virginia Company struggled to maintain its infant colony in the Chesapeake region, it opened the northern part of its American domain to other English settlers. Beginning in the 1610s, James I launched a concerted attack against religious dissidents. He ordered Archbishop William Laud, head of the Church of England, to dismiss ministers with Puritan or Presbyterian leanings. Archbishop Laud pursued this repressive policy with great vigor. He ousted more than 400 preachers from their parishes and forced members of their congregations to accept the Church of England's doctrines and rituals, under penalty of fine or imprisonment.

In response to this persecution, some Puritans separated completely from the Church of England and fled to Leyden in Holland. In 1619, these Separatists secured a patent from the Virginia Company and recruited additional migrants from London. The Separatists, better known as the Pilgrims, arrived at Cape Cod, Massachusetts, in September, 1620, and established Plymouth Colony. Lacking a formal charter from the king, the settlers agreed in the Mayflower Compact to "covenant and combine together into a civil Body Politik." This compact extended into the political sphere the religious ties that bound the Pilgrims into a tight community. More importantly, it signified the assent of the non-Pilgrim migrants to be ruled by the colony's leaders.

Most Puritans remained within the Church of England, but thousands of them migrated to America to create a pure religious society, a "New" England. Settling near the Pilgrims, they used a different means to establish the authority of their new government. In 1629, King Charles II granted a charter to a "Governor and Company" to settle the lands of "the Massachusetts Bay in New England." The king appointed the first governor, but gave the company's shareholders the power to elect future governors and to form a "general court" to conduct other business. The charter, however, did not specify an English city in which this general court was to meet. This error (or carefully planned omission) allowed the Puritans to carry their charter to New England.

Led by John Winthrop, whom they elected governor before their departure, the first non-Separatist Puritan settlers arrived in Massachusetts Bay in 1630. Acting under the authority of the charter, they established a religiously based political order. The migrating shareholders broadened their ranks by enfranchising all adult male members of Puritan congregations in the new colony. Church members elected a governor and a group of magistrates to rule on a day-to-day basis, subject to the ultimate authority of the general court. This action effectively transformed the Massachusetts Bay Company from an economic corporation into a political society.

In contrast to the profit-minded Virginia adventurers, most Puritan migrants saw their mission primarily in spiritual terms. "We shall find that the God of Israel is among us," Winthrop told his fellow shipmates on the *Arbella*. He urged them, with the aid of God, to create a model society in the New World, "that men shall say of succeeding plantations: 'The Lord make it like that of New England.' For we must consider that we shall be as a city upon a hill, the eyes of all people are upon us." As one Puritan minister put it, the migrants were a "saving remnant," sent into the American wilderness to preserve the true principles of Christianity. Their example would stand as a divine beacon to those who remained behind; in God's good time, the migrants might return in triumph to reform England.

The importance of the Puritan mission imbued its leaders with great determination and authority, as did their own social values. As a member of the privileged gentry class of England, Winthrop believed that he had a natural right to rule over the yeomen farmers, artisans, and servants who comprised the majority of the migrants. God had "so disposed the condition of mankind," Winthrop told his companions on the *Arbella,* that "in all times some must be rich, some poor, some high and eminent in power and dignity, others mean and in subjection."

As the elected governor of Massachusetts Bay Colony during most of the

Richard Mather (1596–1669). This Puritan minister migrated to New England in 1635. This woodcut, made shortly after his death, captures Mather's kindly yet intense religiosity and suggests the stern authoritarian character of the Puritan Commonwealth. (*Source: American Antiquarian Society.*)

1630s, Winthrop ruled with an iron hand. He took the lead in expelling all those who questioned the political institutions or his interpretation of Puritan religious doctrines. At his urging, the magistrates expelled Roger Williams from the colony. A strong-willed yet thoughtful clergyman, Williams had served as the minister of the Puritan congregation at Salem. He criticized the refusal of the churches of Massachusetts Bay to separate completely from the Church of England (as the Pilgrims of Plymouth Colony had done). Williams also questioned the religious requirement for voting and the government's arbitrary seizure of the Indians' lands. Following his expulsion, Williams established the town of Providence in Rhode Island, secured a royal charter for the new colony, and implemented many of his beliefs. Other Puritans left Massachusetts Bay to settle in the adjoining colony of Connecticut, which was founded in 1635 and received a royal charter in 1665.

Winthrop took an equally harsh stance in the dramatic trial and expulsion of Anne Hutchinson. This episode was of crucial importance, for it defined the religious doctrines that would prevail in many Puritan congregations. Hutchinson came to Boston with her husband (a merchant) and her children in 1634. She quickly gained remarkable prominence in the community. Her skills as a midwife made her popular among Boston's women. More importantly, her intellectual abilities gained her a following that included both men and women. She began to lead weekly religious meetings in her home, where members of the congregation discussed the previous Sunday's sermon and other topics.

Anne Hutchinson's comments during those meetings soon brought her into conflict with Boston's ministers and magistrates. Like all Puritans, she subscribed to the religious doctrines proposed by John Calvin, the eminent Swiss theologian. She believed that human beings gained salvation only through God's freely given grace. Like her Puritan neighbors, she also held that God chose a select few, before birth, and predestined these "Saints" for heavenly salvation. Thus, a person's good works and pious behavior during life counted for nothing in the divine scheme of redemption.

However, the Puritans' Calvinist theology had many subtle implications. No Puritan minister held that good works could lead to salvation, but most of them emphasized the importance of good behavior as a way of showing respect and obedience to God. They also suggested that good works might be a sign of "election." Anne Hutchinson sharply attacked the Puritan clergy on this issue. She felt that too many Puritans used the outward display of good works to mask the depravity of their souls. In her view, only two Boston ministers — John Cotton, whom she had followed to Massachusetts Bay, and her brother-in-law John Wheelwright — were truly saved and were preaching a "covenant of grace." The other ministers, she claimed, were spiritual backsliders posing under a "covenant of works."

Threatened by Hutchinson's charges, the ministers brought her to court before the magistrates. Governor Winthrop personally led the interrogation. Throughout weeks of discussion, Winthrop and his fellow magistrates confronted Hutchinson with Biblical passages, hoping to convince her of the error of her beliefs. They often heaped scorn upon her because she was a woman. To hold meetings in her house, Winthrop declared, was "not tolerable nor comely in the sight of God nor fitting for your sex." Brushing aside these personal attacks, Hutchinson defended herself brilliantly. She matched her accusers' scriptural evidence passage for passage — and often outmatched them with mental quickness and wit.

Finally, Hutchinson made a slip that undid her. When asked how she knew that one of her assertions was true, she replied that she was so instructed by God, "by the voice of his own spirit to my soul." The judges were astonished by this claim of direct revelation; most Puritans believed that the age of revelation was over and that the Bible was the sole source of God's word. They used this admission to convict Hutchinson of heresy and to banish her from the infant colony. She followed Roger Williams to Rhode Island and, much to the satisfaction of many Massachusetts Bay Puritans, was later killed in an Indian uprising.

Anne Hutchinson's trial clarified both the nature of religious orthodoxy in Puritan New England and the quest for power by the governor and the magistrates. By choice and necessity, the first Puritans initially established an authoritarian political and religious regime. Since nearly 20,000 migrants had arrived in Massachusetts Bay by 1640, Winthrop and the magistrates found it increasingly difficult to impose their authority. A dozen Puritan communities dotted the coastline, and each sent two representatives to the general court. A town was often settled by Puritans from a distinct region of England; these inhabitants wanted to govern themselves in accordance with their traditional social customs. Moreover, various Puritan congregations upheld different interpretations of Calvinist theology, and each group insisted that they, not their ministers or the magistrates, had the right to determine church doctrine.

In 1643, the general court accused Governor Winthrop of exceeding his authority and brought him to trial. Winthrop was humbled by this episode, but sought to turn it to his advantage. He celebrated his acquittal by instructing the court "about the authority of the magistrates and the liberty of the people." "Natural liberty," the governor began, "is common to man with beasts and other creatures. . . . It is a liberty to [do] evil as well as to [do] good." Conversely, Winthrop argued, "civil liberty" stemmed from "politic covenants and constitutions" such as the royal charter of 1629. "This [civil] liberty is maintained and expressed in a way of subjection to authority," the chief magistrate concluded, advising his audience to "quietly and cheerfully submit unto that authority which is set over you, in all the administrations of it, for your own good."

This political philosophy set the tone for government in the Puritan commonwealths of New England for the next half-century, although not quite in the manner Winthrop expected. The power and importance of the general court declined after 1650, and the Puritan-dominated town governments emerged as the prime political institutions. Simultaneously, the lay members of Puritan congregations, the "Saints" who had publicly testified that God's grace had made them members of the Elect, increased their power. They contested the authority of ministers and magistrates like Winthrop and achieved substantial control over church affairs and religious doctrine. Despite this devolving of authority onto local towns and lay congregations, Massachusetts and Connecticut remained strong, unified, and authoritarian Puritan commonwealths. Puritans held most political offices and effectively controlled the religious lives of those colonists (one-half of the population) who were not Puritan Saints or members of their families. In this respect at least, the Puritans' mission into the American wilderness remained a striking success; for better or worse, they had established a coercive holy commonwealth.

North America in 1600

As Puritan Saints and Virginia Company adventurers invaded eastern North America, they encountered people who lived in a very different type of society. The Indians were not organized into a single nation, or even several nations. They had no kings, no hereditary nobility, no specialized groups of merchants, and no subservient peasant class. Instead, native Americans were divided into language groups and, within these, into tribes composed of intermarried clans.

Many distinct tribes inhabited the lands granted by James I to the Virginia Companies of London and Plymouth. The Indians of New England, such as the Abnaki and the Narragansett, spoke various dialects of the Algonquin language. Other Algonquin speakers, such as the Delaware and Susquehanna, occupied lands as far south as the Chesapeake Bay. Muskogean tribes — the Creeks and the Yamassee — lived along the coastline further to the southwest. These tribes varied in numbers from 2000 to 20,000 individuals. Each controlled a defined territory that it defended against all intruders. As an early English settler noted, every village leader "knoweth how far the bounds and limits of his own country extendeth." One boundary line between tribes in Massachusetts ran through a body of water called Chabanakongomuk, which meant "you fish on your side, I fish on my side, nobody fish in the middle."

Tribes usually divided themselves into villages of 200 to 400 persons, each of which was ruled by a *sachem* — a leading man or woman. Unlike English nobles, most sachems exercised limited authority. Their power did not stem from wealth or legal privilege, but rather from personal skills or charisma. Some male sachems rose to power by marrying several wives, thereby connecting themselves with more than one of the tribe's clans (or lineage groups). Other sachems ruled because of their military prowess. In any case, few had great power. "Their authority is most precarious," a French priest noted, "if indeed, that may be called authority to which obedience is in no wise obligatory."

Just as each tribe or village claimed a traditional right to occupy certain territory, so did each clan enjoy the customary use of certain agricultural fields and hunting territories. These land claims did not imply absolute ownership but

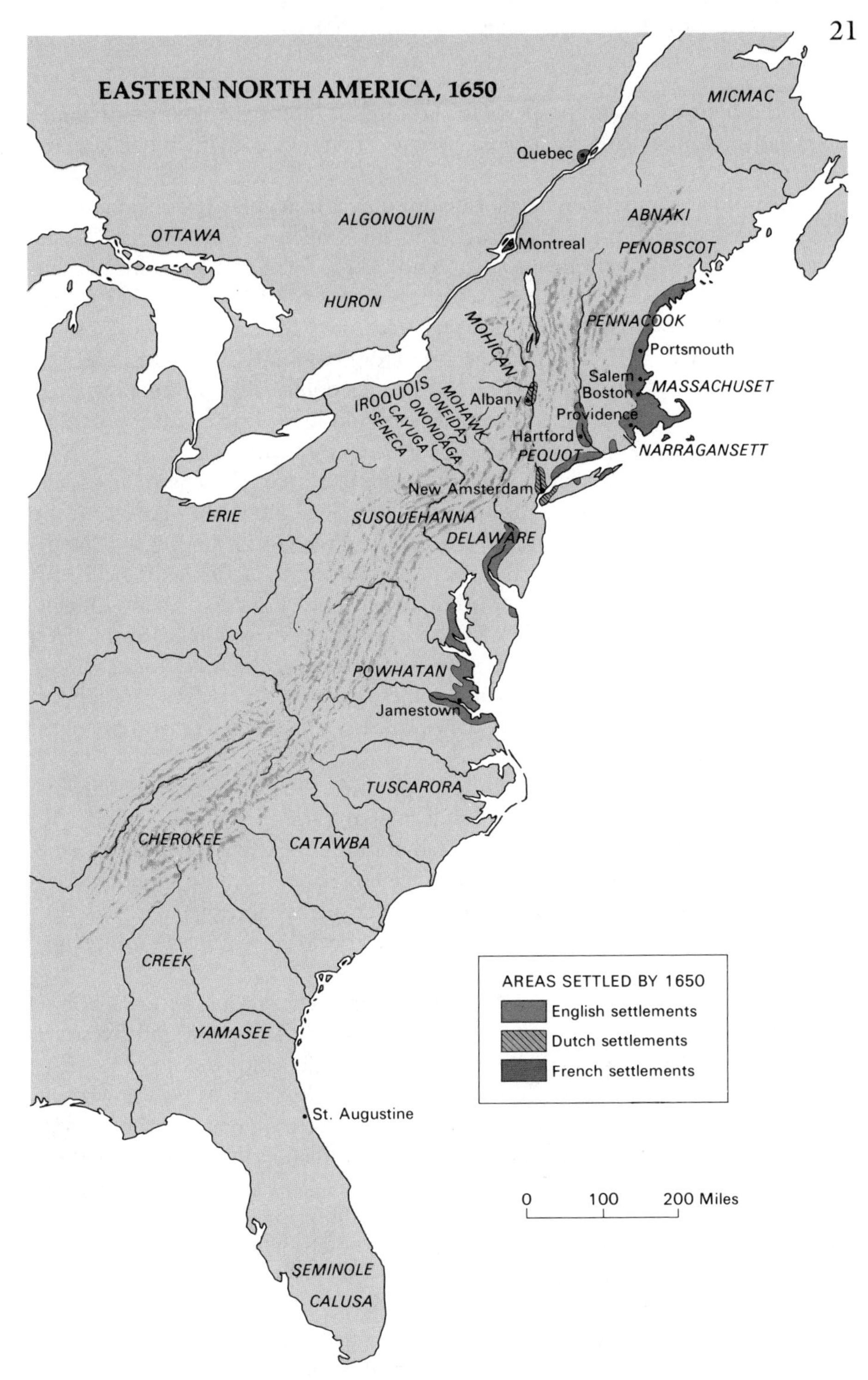

After a half-century of effort, European settlers and fur traders had carved out only a few small enclaves in eastern North America; almost all the land was still controlled by scores of Indian tribes.

simply the right to use the property for farming or hunting. Like English common-field peasants, clans reallocated use-rights among their member families from time to time.

Each Indian family used its allocated resources in regular seasonal patterns. As the winter ended, the clans gathered in their villages. The women carried hoes tipped with flint or bone into the broad cleared fields that surrounded their compact settlements. A Dutch traveler reported that they made "heaps like molehills, each about two and a half feet from the others, which they sow or plant in April with maize, in each heap five or six grains." The corn, and the squash and beans planted with it, provided an adequate diet of vegetable protein. In fact, the food produced by female horticulture (or hoe-culture) provided up to three-fourths of Indian families' caloric requirements.

The gathering of the wild bounty of nature provided tribes with additional food. "From the month of May to the middle of September, they are free from all anxiety about their food," noted an early English visitor among the Maine Indians, "for the cod are upon the coast, and all kinds of fish and shellfish." Interior tribes relied on hunting to obtain their animal protein. Male Indians hunted in bands between October and December; women butchered the deer and bear, cooked and preserved the meat, and processed the hides for clothing. Then, from January to March, Indian families scattered themselves throughout the tribal domain. To survive during the cold season, they relied on stored grains and a few animal kills.

Thus, the Indians changed the location and density of their settlements according to the season of the year. By diversifying their economic activities to include hoe-culture as well as gathering and hunting, the tribes of southern New England and the Chesapeake Bay region created relatively stable societies. And their productive activities changed the character of the landscape; broad cultivated fields of corn, beans, and squash stretched between streams and the dominant forest.

Yet these native cultures were not as economically productive or as politically disciplined as those of the invading Europeans. The grain and livestock economy of England provided an assured supply of food for more people than did the Indians' tribal economy, in part because English men as well as women worked at farming. Kings, nobles, and merchants exerted greater authority and discipline over the English population than the Indian sachems did over their people. England had the social power, economic institutions, and technological skills to dispatch thousands of colonists across 3000 miles of ocean. No tribe, not even the centrally organized Iroquois nation of what is now New York State, possessed such material resources or administrative talents.

The ability of native Americans to resist the invasion of whites was further undermined by disease. Epidemics of measles and smallpox periodically swept through all European societies, taking thousands of lives. These viral infections were even more devastating in America, for Indians lacked inherited or acquired immunities to them. Beginning in the 1580s, visiting English and French fishermen dried their fish on the beaches of New England — and inevitably transmitted their illnesses to the resident tribes. The results were tragic. In 1616, an epidemic virtually wiped out the coastal tribes of Massachusetts. An English colonist reported five years later that the Indians

> died in heapes, as they lay in their houses and the living that were able to shift for themselves would runne away and let them dy, and let there Carkases ly above the ground without burial. . . . And the bones and skulls . . . made such a spectacle . . ., it seemed to mee a new found Golgotha.

Similar epidemics devastated the tribes of other regions. As early as 1585, Arthur Barlowe observed that the Algonquin tribes in Carolina were "marvelously wasted, and in some places, the Countrey left desolate." In 1608, shortly after they had arrived at Jamestown, the Virginia colonists noted that a "strange mortalitie" took the lives of " a great part of the people" of the Accomacke tribe. Everywhere the Europeans landed, the story was the same. A Dutch chronicler reported from New Netherlands in 1656 that

> the Indians affirm, that before the arrival of the Christians, and before the small pox broke out amongst them, they were ten times as numerous as they now are, and that their population had been melted down by this disease, whereof nine-tenths of them have died.

As one historian has noted, the first English settlers found not a "virgin" land but rather a "widowed" one.

The Collision of Cultures

According to its charter, the Virginia Company was to bring Christianity and civilization to the Indian inhabitants of North America, "to cover their naked misery, with civil use of food and clothing, and to train them by gentle means to those manual arts and skill, which they so much affect and do admire to see in us." From the first days of settlement, however, the English looked upon the natives as objects of exploitation and obstacles to expansion. By one strategem or another, they tried to bully the Indians out of their food and their lands.

Like all conquerors, the English invaders clothed their naked aggression in high-minded justifications. In his pamphlet titled "The Lawfulness of Removing Out of England into the Parts of America," Robert Cushman, one of the Pilgrims who settled in Plymouth Colony, suggested that Indians "do but run over the grass, as do the foxes and wild beasts." "The Indians are not able to make use of one fourth part of the land," argued a Puritan migrant to Massachusetts Bay. Therefore, minister John Cotton of Boston concluded, "in a vacant soyle, he that taketh possession of it, and bestoweth culture and husbandry upon it, his Right it is."

These justifications for seizing Indian lands had little basis in fact. Indians did "change their habitations from place to place," as Cushman pointed out, but they also lived for much of the year in settled agricultural villages and main-

An Indian Village in 1585. During the summer, most eastern seaboard tribes lived in compact villages set in the midst of cultivated fields. Each wigwam housed several families who were members of the same clan or lineage group. (*Source: The Granger Collection.*)

tained control over clearly defined territories. What counted in the end was not factual accuracy but political policy and military power. Roger Williams, the influential Puritan minister, suggested that the king of England had committed an "injustice, in giving the country to his English subjects, which belonged to the Native Indians." Most Puritans brushed aside Williams's arguments. Their religious faith led them to believe that the Indians worshipped "False Gods and Devils," that Indian *powwows* (or religious leaders) were witches, and that the Indians themselves "serve the Devil and are led by him." Why should creatures of Satan deprive God's Saints of land?

Among colonial leaders, only William Penn, the proprietor of Pennsylvania (founded in 1681), adopted Williams's enlightened perspective. Penn's Quaker religious doctrines determined his actions; these doctrines stressed the presence of an "Inner Light" in all people. Penn therefore treated Indians as near-equals and purchased their land by negotiated treaties. Nearly all the other colonies claimed title to the land by virtue of their royal charters. "All lands in this government are holden of the King of Great Britain as the lord of the fee," the Connecticut General Court declared in a legal case in 1717, and "no title . . .

can accrue by any purchase made of Indians on pretense of their being native proprietors thereof."

English desire for Indian land placed the two peoples in conflict, and day-to-day contact increased the level of tension. The Virginia adventurers arrived in America with conflicting images of the native inhabitants. Some prior descriptions, written by the Spanish invaders of Central America and early explorers of the North American coast, pictured Indians as "a loving people without covetousness — who become entirely our friends." Other early accounts conveyed a harsher picture. They described Indians as "Brutish savages" who would eat any "flesh or fishe . . . (smell it ever so filthily)."

The course of events gave credence to each of these images. Indians led by Powhatan, the leading sachem of the many small tribes of the Chesapeake Bay area, welcomed the English invaders as potential allies against more distant tribes. "In our extremity," Captain John Smith reported in 1607, "the Indians brought us Corne . . . when we rather expected . . . they would destroy us." Two years later Powhatan cut off trade with the newcomers, having become convinced (he told Smith), "your coming is . . . to take my people and possess my country." Smith responded by raiding Indian villages for corn. "You promised to freight my ship [with corn] ere I departed, and so you shall," the blunt English adventurer warned the sachem of the Pamunkey tribe, "or I mean to load her with your dead carcasses." Soon sporadic warfare broke out between the two peoples. By 1614, the English had forced Powhatan to sign a humiliating peace treaty, in which he acceded to the invaders' demands for corn and land.

An uncertain peace soon gave way to outright war. The spread of tobacco production and the Virginia Company's new land policy brought an influx of new colonists — more than 1000 just between 1619 and 1622. To prevent the destruction of his people through the spread of European settlers and diseases, the sachem Opechancanough organized the Chesapeake tribes into a military alliance. They attacked the English settlements in March, 1622, in retaliation for the murder of a religious prophet. Surprising outlying plantations, Opechancanough's forces killed over 350 English settlers, nearly one-third of the white population.

This stunning assault brought the downfall of the Virginia Company. In London, the company's directors strained their resources to the limit by quickly dispatching 3000 new settlers to the threatened colony. Another consequence of the disaster was that it deprived the company of royal support. Accusing the company of gross mismanagement, James I transformed Virginia into a royal colony in 1624. At the same time, English settlers took military revenge, beginning "a perpetual war without peace or truce." "Our first work is the expulsion of the Savages . . . ," vowed Governor Francis Wyatt, "for it is infinitely better to have no heathen among us." Another Virginian exclaimed:

> We . . . are now set at liberty by the treacherous violence of the Sauvages . . . So that we . . . may now by right of Warre, and law of Nations, invade the

> Country, and destroy them who sought to destroy us; whereby wee shall enjoy their cultivated places.

The colonists dispatched expeditions to destroy Indian villages and crops and to "root [them] out from being any longer a people."

The arrival of thousands of new settlers sealed the fate of the Chesapeake tribes. The steady expansion of tobacco fields pushed the survivors out of the tidewater area. When the desperate tribes staged another uprising in 1644, there were more than 10,000 migrants to oppose them. To prevent further violence, the Virginia government concluded a formal peace treaty with the Indians in 1646. This document guaranteed certain lands to the displaced tribes; in return, the Indians pledged an annual tribute of beaver skins and military assistance in the event of attack by other tribes. Within another generation, sporadic fighting and European diseases had nearly destroyed the natives of the Chesapeake region. An English census in Virginia in 1669 revealed only 11 tribes and 2000 Indians, a far cry from the 28 tribes and 20,000 natives described by Captain John Smith in 1608.

A similar fate befell the Indians of New England. The arrival of the Pilgrims in 1620 resulted in new epidemics among the already devastated tribes of southern New England. "The chief sachem himself now died," Pilgrim leader William Bradford reported, "and almost all his friends and kindred." Those that are left, Robert Cushman somewhat gloatingly announced, "have their courage much abated, and their countenance is dejected, and they seem as a people affrighted." For many Pilgrims and Puritans, this "sweeping away [of] great multitudes of natives" seemed a sign of God's intervention to "make room for us there."

Taking advantage of the weakened condition of the Indians, Pilgrim Miles Standish led several attacks against the Massachusetts tribe. To frighten the Indians into submission, Standish undertook a campaign of calculated brutality. Following one incident, for example, he impaled the head of the sachem on the Plymouth fort. Thereafter, an observer noted, the Indians called the colonists "Wotowquenarge, which in their language signifieth stabbers or Cut-throats." Deterred by this violence (and further weakened by a terrible smallpox epidemic in 1633–1634), the tribes offered little initial resistance to the great wave of Puritans who migrated to New England.

Competition for land and furs soon resulted in full-scale warfare between the invaders and the resident tribes. Most Puritans established farming communities, first in Massachusetts Bay and then in Connecticut and Rhode Island. Seeking an exportable commodity, they expanded existing Indian trading networks. Puritan farmers raised Indian corn, and Puritan merchants traded it for the *wampum* (strings of carefully drilled and polished seashell beads) produced by the tribes along Long Island Sound. Then the merchants exchanged the wampum for tanned beaver furs produced by tribes in northern New England.

To exclude the Dutch merchants of New Netherlands from this lucrative export trade in furs, the Massachusetts Bay and Plymouth governments formed an alliance with the Narragansett tribe. A joint English-Narragansett military force attacked the strong Pequot tribe, which had continued to trade with the Dutch. In May, 1637, the English-led forces surrounded and burned the main Pequot village, killing many members of the tribe and selling the rest into slavery in the West Indies. "It was a fearful sight to see them thus frying in the fire . . . ," William Bradford confessed, "but the victory seemed a sacrifice . . . over so proud and insulting an enemy."

The destruction of the Pequots established English dominance in southern New England. The conquest allowed Puritan merchants to seize control of the trade in wampum and furs. From his trading post at Springfield, on the Connecticut River, a single merchant, John Pynchon, secured 9000 beaver furs from Indian trappers between 1652 and 1658. This massive export trade severely depleted the beaver (and deer) population of New England, while temporarily improving some material aspects of Indian life. Indians eagerly traded with Pynchon and other European merchants for iron pots and hoes, steel knives, and guns — for these goods were more durable than flint tools and clay vessels. Similarly, many tribes exchanged their traditional garments of skins and furs for "a kind of cloth, called duffels, or trucking cloth, about a yard and a half wide."

The fur trade accelerated the decline of Indian culture. The use of European goods undermined traditional handicrafts in many tribes, increasing Indian dependence on imported goods. Tribes began to fight with each other over rights to the dwindling supply of fur-bearing animals. Even the victors in this struggle did not really win, for they often received rum in exchange for their furs. Overconsumption of alcohol killed scores of Indians and sapped the morale and self-confidence of those who did not die. The combined effects of warfare, disease, and the fur trade virtually destroyed the tribes of southern New England. Their numbers declined precipitously from perhaps 120,000 in 1570, to 70,000 in 1620, to 12,000 in 1670.

As tribal cultures disintegrated, Puritan ministers stepped up their activities. To redeem the native Americans from "Devill worship" and speed their conversion to Protestant Christianity, John Eliot translated the Bible into the Algonquin language. Beginning in 1651, he encouraged Indians to settle in "Praying Villages." As a prerequisite to conversion, Eliot and other Puritan ministers required Indians to become "civilized" by abandoning their clan-based longhouses and their traditional games, customs, and dress. They had to conform to the conquerors' norms, such as by wearing "their *haire* comely, as the *English* do." Regardless of their physical appearance, few Indians met the rigorous spiritual tests required for Puritan conversion, and those that did were still not fully accepted into English congregations.

In 1650, as in 1600, the English settlers and the native Americans lived in distinct religious and cultural worlds. Yet their fortunes had been intertwined for two generations, and both groups had been deeply affected by the contact. English settlers depended on Indian corn for most of their food and on tobacco and the Indian fur trade for most of their exports and wealth. The native Americans were greatly reduced in number, as a result of European diseases and warfare, but the survivors farmed, hunted, and clothed themselves with English goods.

The English invasion had succeeded, but native Americans had a definite impact on the political and social (as well as the economic) evolution of the developing colonies. The Indian uprising of 1622 destroyed the Virginia Company and significantly influenced the subsequent course of politics. Planters who condemned Governor George Yeardley for standing around as "a cypher whilst the Indians stood ripping open our guts" united to force the ouster of Governor Harvey, who favored a negotiated agreement with the Indian tribes. Their antipathy toward the "Savage Devills" brought out the worst aspects of the Puritans' character, and the danger of Indian attack heightened the authoritarian nature of their political regimes. Forced by desire and circumstances to become conquerors, the English were unable simply to transplant their existing institutions and values to America.

2

Internal Conflict and Imperial Control, 1650–1700

By 1650, after a generation of settlement, more than 50,000 English colonists had a firm foothold in North America. They had established communities and institutions that served their purposes, first for survival and then for more extensive development. Although the early American colonists shared a common English heritage, the new society they created was quite different from that of their native country. Their settlements were neither designed nor directly controlled by officials in England.

Equally important, the settlements were also quite different from one another. The agrarian villages of New England and the plantations of the Chesapeake region represented decidedly distinct forms of social organization, based on different assumptions and serving different functions. Both forms, however, had a critical influence on the development of colonial American society. In 1650, the two leading New England colonies (Massachusetts and Connecticut) and the two leading Chesapeake area colonies (Virginia and Maryland) together contained over 85 percent of the colonial population. By 1700, when the English population of all the colonies had surpassed 250,000, those four colonies still accounted for two-thirds of the total. Thus, the nature of those colonies defined the context of everyday life for the vast majority of Anglo-Americans in the seventeenth century.

In both regions, social relations changed subtly yet significantly in the second half of the seventeenth century. As the English population expanded in size, it spread into the interior. White settlers continued (and essentially completed) their conquest of the native tribes. Conflict did not end there. Colonists also

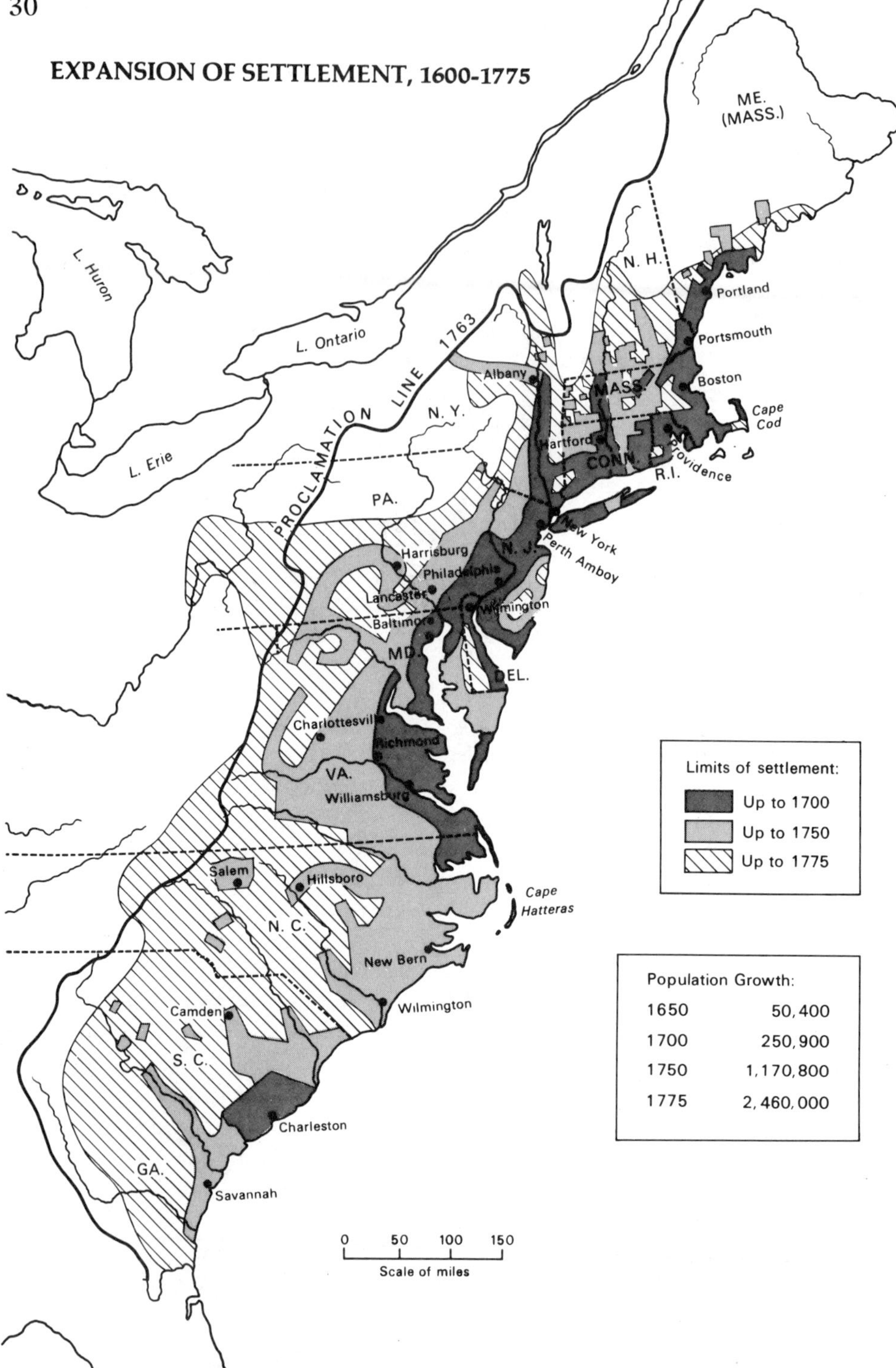

Europeans first settled near the coast, usually along navigable rivers or within fifty miles of the ocean. After 1700, the rapidly expanding white (and slave) population penetrated further and further into the interior, displacing native American tribes.

turned on one another as they asserted control over the land and its resources. Many people who felt themselves at an unfair disadvantage struck out at well-established, more prosperous settlers. Although the resulting conflicts were not as violent or vicious as those with the Indians, they nevertheless revealed deep tensions in both New England and Chesapeake society.

England's attempt to expand its control over the economic and political development of its colonial possessions created yet another source of conflict. In the first half-century of settlement, English officials dealt with North American issues on an *ad hoc* basis. Beginning in 1651, however, English kings and ministers devised a series of economic and political policies designed to regulate colonial trade. Colonists accepted some of the restrictions imposed by the Navigation Acts, but ignored or evaded others. Then, threatened by attacks on their political and religious institutions, settlers in three colonies rose in revolt against England in 1689. However, the American colonists who challenged royal authority in 1689 were localist rebels, not ideological revolutionaries. They wanted autonomy within the English colonial empire, not an independent existence. By 1700, they had largely achieved their goal.

The Navigation Acts

English government underwent dramatic political changes in the middle of the seventeenth century, but official policy toward the colonies remained constant — to control colonial trade. Beginning in 1641, England was torn by civil war. The invasion of an army of Scottish Presbyterians sparked an uprising of English Puritans. Parliamentary leaders deposed King Charles I in 1642 and executed him in 1649. Having achieved substantial representation in Parliament, the Puritan merchants of London secured passage of a Navigation Ordinance in 1651. This act required colonial products to be shipped in English- or colonial-owned vessels.

Following the restoration of the Stuart monarchy in 1660, King Charles II (1660–1685) approved a new and far-reaching Navigation Act. This legislation prohibited foreign vessels from participating in the English colonial trade. In addition, the act stipulated that all colonial "Sugars, Tobacco, Cotton . . . or . . . dying wood" had to be shipped to England, and it placed import duties on most of these enumerated commodities. Then, in 1663, a Staple Act required that European exports to the colonies be routed through England. This act further increased royal customs revenues while providing additional income for English merchants. Finally, in 1673, a Plantation Duty closed loopholes that the colonists had used to export tobacco and other products directly to European markets.

In less than twenty-five years, the English government had pragmatically created a comprehensive system of trade regulations. The purpose of this legislation, the royal treasurer explained, was "to turn the course of a trade rather than to raise any considerable revenue to his Majesty." In fact, these Navigation Acts benefited the monarchy financially in a variety of ways. As early as the 1660s, the duties on tobacco from Maryland and Virginia amounted to 25 percent of English customs revenues and 5 percent of the home government's entire income. In 1675, these tobacco levies yielded £100,000 a year, and in 1700, £400,000.

The economic and political benefits of the Navigation Acts did not end there. First, they cemented the longstanding alliance between the monarch and the English merchant community. By granting a monopoly over colonial trade to his merchant allies, Charles II secured both the political support of the merchants and a ready source of financing for wars and other royal expenditures. Second, the acts ensured English self-sufficiency with regard to important semitropical crops, such as sugar, tobacco, and indigo, and guaranteed English manufacturers a virtual monopoly over colonial markets. Finally, the Navigation Acts enhanced England's world position. Most political economists of the time agreed that trade formed the basis of national wealth. With considerable justification, they argued that a strong merchant marine served as a "nursery for seamen" and was the basis for national power.

The Navigation Acts led directly to three naval wars with the powerful Dutch Republic. For more than a century, Dutch merchants had dominated European trade. With trading posts in India, Indonesia, and China, the Dutch controlled the commerce of the East Indies. The Dutch African Company carried thousands of slaves to the western hemisphere. The Dutch West Indies Company controlled sugar plantations in Brazil and established fur-trading posts in its North American mainland colony of New Netherlands. Attempting to limit Dutch commercial expansion, England battled Dutch ships in European waters between 1651 and 1654, seized New Netherlands in 1664, and defended the new colony of New York when fighting broke out again in 1674.

The American colonists greeted these measures with defiance rather than enthusiasm. Governor William Berkeley of Virginia protested that royal duties doubled the price of tobacco, cutting English demand and planters' profits. He had harsh words as well for the Navigation Acts' exclusion of Dutch traders. The "forty thousand people" in the Chesapeake region, he wrote in 1662, would "be impoverished to enrich little more than forty [London] merchants, who being the only buyers of our tobacco, give us what they please for it."

In New England, Puritan merchants and magistrates welcomed the conquest of New Netherlands, for it eliminated the Dutch as rivals in the fur trade. At the same time, Puritan merchants continued to trade directly with European nations and refused to obey the Navigation Acts. "The laws of England are bounded within the four seas," the Massachusetts General Court argued, "and do not reach America." If the colonies of New England were not restrained, warned an English official, they would soon grow to "be mighty and powerful and not at all careful of their dependence upon Old England."

Local Autonomy in Puritan New England

The roots of resistance to external authority ran deep in the culture of early New England. English Puritans spoke of their migration to the New World as a mission for the spiritual, if not economic, welfare of England. After their arrival, however, they soon focused on their own concerns. With their corporate charter securely in their possession, the original migrants established a self-governing commonwealth in Massachusetts. Moreover, the social and religious imperatives of Puritanism led many settlers to turn inward among themselves. Throughout the first century of settlement, colonists lived a life of relative independence, isolated from the homeland and heritage they had left behind.

Unlike the widely dispersed planters in the Chesapeake area, New England settlers grouped themselves together in small villages. The pattern of village life was not the same in all New England towns. The earliest seventeenth-century towns varied in area (50 to 100 square miles) and population (20 to 50 families). They also had different systems of land distribution. Many migrants came in groups from a particular county or community in England, and their regional background had a great influence on the nature of their New World settlements.

Yet a number of shared traits greatly overshadowed these differences. The majority of New England settlers were Puritans, following a common religion within the limited social space defined by the town. The ethos of New England town life encompassed both the individual and the family within a broader social structure that created — or reinforced — ties of mutual commitment and cooperation. John Winthrop emphasized to his fellow Puritans that

> we must be knit together in this work as one man. . . . We must delight in each other, make others' condition our own, rejoice together, mourn together, labor and suffer together: always having before our eyes our commission and community in the work, our community as members of the same body.

Winthrop's biblical image of a "city upon a hill" gave a visual embodiment to the form of this collective endeavor. For more than two centuries, the New England town remained the most immediate source of personal identity and social order. The Puritans' commitment to living in close community provided the common basis for their political culture.

The most prominent institution of communal self-government was the church. When the first settlers of Springfield, Massachusetts, drafted their "Articles of Agreement" in 1636, they made clear their concern for religion:

> We intend by God's grace, as soon as we can, with all convenient speed, to procure some Godly and faithful minister with whom we purpose to join in church covenant and walk in all the ways of Christ.

In Springfield, as in most New England towns, the settlers established only one church, which embraced all but the most reprehensible residents. Indeed, the location of the meetinghouse at the center of town aptly symbolized the centrality of religion in the community's life.

The Puritan church stood as a self-contained congregation of believers who generally adhered to a common doctrine, accepted a common discipline, and abhorred any form of deviation. They hired their own minister and expected him to give them spiritual and temporal guidance. Twice a week — once on Thursday for a lecture and again on Sunday for a sermon and extended service — the entire community assembled at the meetinghouse to hear "the Word" preached by the minister. Repeated religious instruction represented the most important and effective means of reinforcing the personal morality and collective ethos of Puritan society. People might learn these lessons in private through catechisms and Bible study, but they also heard them propounded regularly and publicly from the pulpit.

Religious services had a social significance that went far beyond moral instruction. In the comparatively closed world of an agrarian village, regular collective activities reinforced communal identity. People in New England towns looked forward each year to a number of events marking the passage of the seasons and the generations — town meetings, militia musters, harvests, house raisings, weddings, baptisms, and funerals. But no other gathering had the same degree of regularity and inclusiveness as religious meetings, and none provided quite the same degree of face-to-face familiarity and social cohesion. In church, people actually saw the whole community assembled in one place for a common purpose.

Church services also reflected spiritual and social distinctions within the community. Most towns assigned seats in the meetinghouse according to age, wealth, and status, thereby reinforcing physically a distinction between old and young, prosperous and poor, and in some cases male and female. Moreover, until the early part of the eighteenth century, most churches reserved the sacrament of communion for those select "Saints" who had experienced — and could give public testimony of — the working of salvation in their souls. Thus, elements of a spiritual and social exclusivity coexisted with a broader communal inclusiveness: people learned their "place" in the community every time they went to church.

The meetinghouse was also the site of another important communal activity, the town meeting. The voters of the community met at least once a year, usually in March, to discuss and resolve a variety of community concerns. The town meeting was not a complex institution. Rather, it addressed the limited needs of an agrarian village: selecting town officers and functionaries, dividing land, laying out roads, marking boundaries, repairing fences, and regulating the grazing of livestock or the behavior of people. Moreover, the town meeting was less a forum for engaging in debate than a means of reaching a consensus and expressing agreement. Voicing common values, ministers and selectmen urged townspeople to put aside individual interests for the welfare of the community

Interior of the First Parish ("Old Ship") Meetinghouse, Hingham, Massachusetts. The meetinghouse was the site of both religious services and political meetings in the New England town. The repeated gathering of the townspeople in this enclosed space reinforced the bonds of community. (*Source: Imants Ansbergs*)

as a whole. In 1645, the inhabitants of Dorchester, Massachusetts, admitted that they were "heartily sorry for and ashamed of" the conflicts that had recently beset their town meeting, and they established new bylaws requiring that "all men shall attend to what is propounded by the [selectmen] . . . avoiding all janglings by two or three in several companies, as also [avoiding] to speak unorderly or unseasonably." The political emphasis on harmony and unity reflected the religious emphasis on community. Like the church, the town meeting was intended to be both a source and a symbol of the townspeople's common identity and commitment.

Also like the church, the town meeting reflected the balance, or perhaps the tension, between exclusivity and inclusiveness. Decisions in the town meeting had to be accepted by everyone, but were not made by everyone. In any New England town, the majority of the inhabitants — women, young people, black slaves, Indians, and even adult white men who did not own land — could not formally take part in politics. Some of them, especially women and economically dependent sons, might express their opinions and assert their influence in private, but the decision-making power remained in the hands of property-owning white males over the age of 21. As heads of households, husbands spoke for —in fact, most often defined — the interests of their wives, children, and servants. In practice, many towns allowed almost all male heads of households to take part in local politics, even if they did not exactly meet the property requirements. In the town meeting, as in the church, the creation of collective identity required collective involvement.

Moreover, most men in seventeenth-century New England towns had a reasonably good prospect of holding a town office at some point in their lives. In Sudbury, Massachusetts, thirty-nine men — over half of the town's land grantees — held the office of selectman between 1639 and 1655; during that same period, almost every land grantee held some town office. In Sudbury, as in most New England towns, the position of selectman was usually occupied by men who had attained greater wealth and age, and presumably greater experience and expertise, than their fellow townsmen. Younger or less wealthy men assumed less exalted or less desirable positions, such as fence viewer or hogreeve. In any case, the men who shared the community's resources also shared its responsibilities. In general, men in New England towns enjoyed a greater degree of political participation than they or their ancestors had known in England and than most men experienced in the Chesapeake region.

Agrarian Rhythms and Village Life

The people of the New England towns were governed not just by their own communal institutions, but also by natural temporal rhythms. The changing seasons had a profound effect on all aspects of life within the community and the family. Even the calendar reflected the cycles of an agrarian society. The chronological year ended on December 31, but the new year did not really begin until March 15 (and until then the date was written, for example, as 1650/51). The reason for delaying the new year until March was clear. The middle of March signaled the beginning of the agricultural cycle in the northern hemisphere. The sun was in the sky for a longer period each day, bringing the change in weather that was essential to the economic viability of the town.

With the passing of the long dark winter, the hard work of the farm commenced. Fields were drained, plowed, and planted. Sheep were washed and shorn of their thick wool. In June, the first crop of hay was cut, dried in the fields, and stored for the coming winter's feed. The hard work of haymaking stimulated thirst, and this too had its effect on the economy. "My demand with my country customers is so large now in hay time," one Boston brewer complained to a friend, "that I cannot distill enough for them." The most strenuous (and most productive) work — plowing, sowing, haying, and harvesting — consumed only seventy or eighty days a year, however. When the harvest was over in the fall, the level of activity slowly began to subside, and somewhat less arduous (if no less necessary) tasks began to occupy the greater part of the family's time. October was the time to pick fruit and either dry it or distill it into cider. November was the traditional butchering time, when the year's supply of pork, beef, and lamb was salted and smoked. During the late fall and winter months, members of the family engaged in a variety of indoor tasks, ranging from repairing implements to making clothes. For women especially, food preparation and child care remained constant daily activities, no matter what the season.

The creation of human life also conformed to the cyclical pattern of the wider environment. As Chart 2.1 indicates, the number of successful conceptions (as measured by births nine months later) rose steadily from a low point in February and March until it reached a high point in June. It then declined gradually until September, after which it increased during the autumn months before regressing yet again during the depths of winter. This periodic increase and decrease in conception may have been the result of variation in the frequency of intercourse, in female fertility, or in the incidence of miscarriage (as a result of monthly changes in nutrition and disease). Although the precise cause is not known, the existence of this rhythmic pattern is not in doubt. Year after year, in village after village, the conception cycle followed the same regular schedule. It stands as pervasive and profound evidence of the influence of the annual cycle of nature on these agrarian communities.

CHART 2.1
Monthly Frequency of Successful Conceptions

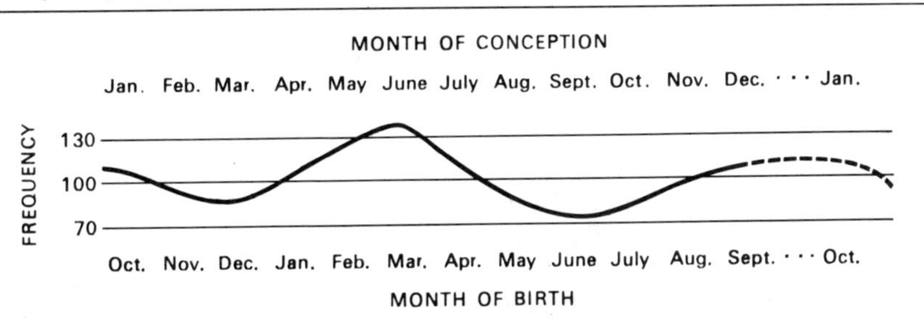

SOURCE: Kenneth A. Lockridge, "The Conception Cycle as a Tool for Historical Analysis," paper presented at the SUNY-Stony Brook Conference on Social History, June, 1969.

The average number of births each month in Dedham, Massachusetts, and in many other farming communities of seventeenth- and eighteenth-century America varied markedly. Successful conceptions (as measured by births nine months later) occurred more frequently in June than in any other month. If successful conceptions had occurred with equal frequency in all months, this chart would show a horizontal line at 100.

The Implications of Population Growth

Time is not just cyclical, however; it is also linear. Despite the tendency — even the desire — of the New England settlers to live a stable and static existence, the life of the Puritan town was inevitably marked by a process of change. Some of the changes that took place were seemingly natural, almost imperceptible results of decisions made within individual households. Yet, when added together over a period of years, the combined choices of many families transformed the nature of the whole community in ways that were significant and sometimes rather dramatic. Moreover, as it was repeated in town after town, the process of local change contributed to the transformation of New England society.

Family size had an especially profound effect on town development. Puritan ministers preached the biblical injunction to "be fruitful and multiply," and their congregations obeyed it with remarkable success. The first settlers of New England came from an Old World background in which the mortality rate was quite high. In their historical experience, an average of five births per marriage had been necessary to maintain even a low rate of population growth. Their New World environment had ecological and epidemiological conditions that were significantly better than those in England. Still, in the seemingly vast and sparsely populated landscape of New England, they adhered to old patterns of high fertility in order to expand a small and vulnerable population.

To some extent, the New England settlers exceeded those Old World demographics. In seventeenth-century Massachusetts towns like Hingham, Andover, or Ipswich, the average age at first marriage for women was between 21 and 22 years, significantly lower than the average age in contemporary English towns like Canterbury (24 years) or Colyton (29.6 years). A woman who married in her early twenties could expect to be pregnant on a fairly regular basis, usually at intervals of twenty-four to thirty months, for the next quarter-century. During the period of breast-feeding, lactation created hormonal changes that provided a natural form of birth control, but once an infant was weaned, the woman was likely to become pregnant again.

Given the comparative healthfulness of the New England environment, most pregnancies resulted in live births. Moreover, most children lived beyond infancy; in Andover and Plymouth, Massachusetts, nearly nine out of every ten children born before 1700 survived to reach the age of 20. By the time an average married woman reached menopause in her early to middle forties, she and her husband would be likely to have had six to eight children. When these offspring grew to maturity, married, and began families of their own, they did more than continue the family line. They also contributed to an ever increasing population. Between 1650 and 1700, the population of the New England colonies grew from just under 23,000 to over 93,000.

The rapid expansion of English settlements transformed both the countryside and the lives of the region's inhabitants. Indians experienced the most dramatic — and traumatic — effects of the growth of the white population. The spread of English settlements across the land steadily encroached on tribal territories. In 1675, a confederation of tribes rose up in rebellion. Under the direction of Metacom, a Wampanoag leader whom the English called King Philip, the Indians of New England staged widespread attacks on English settlements. During the summer, they struck towns on the outskirts of Plymouth Colony and along the Connecticut River; by winter, Indian raiders had penetrated to within twenty miles of Boston. Bitter warfare continued into the spring of 1676. Indian warriors destroyed twelve English towns and attacked forty others, killing nearly 2000 whites. Indian casualties were twice as high as that, and the English destroyed many Indian villages in retaliation. The Indian confederation gradually fell apart. The English finally surrounded Metacom and forced him to surrender; then they executed him and carried his head back to Boston. With the death of Metacom and the destruction of his confederation, the Puritan conquest of New England's native peoples was virtually complete. In the course of the seventeenth century, thousands of Indians had been subdued, destroyed, or driven from the region.

Population growth also strained the spiritual and economic resources of the settlers. Both of the main institutions of New England life — the church and the town — adjusted to accommodate greater numbers of people. In neither case was the adjustment easy or altogether successful.

By the middle of the seventeenth century, many ministers noted a decline in the spiritual intensity and social inclusiveness of religious observance. Part of this decline was due to the growing number of non-Puritan migrants arriving in New England each year. Of much greater concern, however, was the spiritual state of the "rising generation," the children born and raised in Puritan communities. Originally, only the children of Saints — those Puritans who had experienced conversion and were among the spiritually elect — could be baptized. Ministers began to realize, however, that the number of Saints, and therefore the number of Saints' children, represented a decreasing proportion of the population. Accordingly, in 1662, a synod of Puritan ministers proposed a major change in the requirements for baptism. In order to extend the sacraments (and the authority) of the church to more people, the synod created a so-called Halfway Covenant for parents who had been baptized but had not yet experienced conversion. As long as these parents accepted the doctrine and discipline of the church, they could have their children baptized. Thus, more people could enjoy a strong sacramental identity with the church.

At the end of the seventeenth century, Solomon Stoddard, the minister in Northampton, Massachusetts, went one step further. He offered communion to virtually everyone, or at least to those whose lives were not scandalous. Like baptism, communion was a sacrament generally reserved for the converted, and therefore was a clear sign of the distinction between the saved and unsaved. Stoddard argued that communion could be a "converting ordinance" — a means of helping people experience the work of salvation. Like the proponents of the Halfway Covenant, he hoped that if the church could extend its sacraments to more people, it could also increase its influence over them. More than any other Puritan minister, Stoddard emphasized equality over exclusivity.

Both of these measures generated heated controversy among the New England clergy, and most congregations rejected the measures at first. The deviation from strict sacramental purity, they thought, was a sure sign of spiritual decline. But gradually the majority of churches accepted the Halfway Covenant (if not Stoddard's open communion) as a necessary adaptation to a growing and changing population. At the very least, the covenant offered a way to make sure that those who came to the meetinghouse did so with a clearer religious identity, and perhaps even greater religious intensity.

Population growth created economic and political concerns as well. The desire of parents to keep their children close at home and still provide an adequate landed estate for them often strained the resources of the family. To provide an inheritance for several sons, many fathers acquired property several miles away from the village center. Some then consolidated their own holdings on the outskirts of town and moved their families there; others simply gave parcels of outlying property to their sons.

In many cases, the inheritance strategies designed to maintain family continuity in the community contributed to the instability of the town as a whole. Outlying areas came to be inhabited by a growing number of people who had moved out from the center of town. They were joined by newcomers moving to the community for the first time. Eventually, the population of "outlivers" became great enough to support independent institutions and an autonomous community. In town after town, this process repeated itself. Outlivers began to complain of the disadvantages they suffered as a result of their distance from the center of communal activity, and they demanded institutions of their own. The inhabitants of an outlying part of Salem expressed this discontent in petitioning for separation from the parent town in 1667:

> Some of us live ten miles, some eight or nine; the nearest are at least five miles from Salem meeting-house (upon the road) — and then 'tis nearly a mile farther to the sentry-place. . . . And yet [we are] not excused from paying our part to all charges, both ecclesiastical and civil, besides the maintenance of our families [in] these hard times, when the hand of God is heavy upon the husbandman.

However logical in terms of geographical reality, these requests usually kindled hot political fires. The creation of new administrative and religious entities lowered the prestige and increased the tax burden of the original settlements. Therefore, the inhabitants of the town center usually resisted the outlivers' push for separation. But the outlivers still insisted that they needed independence. The desire for separation often led to protracted conflicts that went against the Puritan ideals of unity and harmony. A group of outlivers in Dedham even tried to take control of the town meeting by threatening the moderator with their muskets.

On another level, these conflicts over town divisions reflected an enduring commitment to the communal ideals of Puritan society. Outlivers sought not so much to destroy an old community as to create (or recreate) a new one. They attempted to break out of the linear pattern of communal growth by returning full circle to the small village environment they had once known, and still valued.

Events in Salem revealed the intensity these sentiments sometimes generated. After first petitioning for separate status in 1667, outlivers in Salem had to wait five years before the Salem town meeting granted them a limited degree of autonomy. The inhabitants of Salem Village (as the new section was called) received permission to form a church and hire a minister; the church taxes they had formerly paid to Salem would go for the support of their own church. The town meeting made it clear, however, that Salem Village was not politically independent. The village inhabitants still paid their civil taxes to Salem Town (as the original settlement became known), and they remained under the jurisdiction of that town meeting.

The tensions and animosities surrounding the separation of Salem Village from Salem Town festered for the next twenty years, creating divisions within the village itself. Some village residents, especially those that lived close to Salem Town, were content with limited independence. They found it comparatively easy to remain involved in the political and economic life of the town, which was becoming an increasingly busy and prosperous seaport. Village residents who were farther away from the town looked with disapproval (and perhaps considerable jealousy) at the growing wealth and worldliness of the port. These plain farmers became increasingly hostile toward both the inhabitants of the town and many of their own neighbors in the village. In 1687, a committee of arbitrators called in to resolve a local dispute chided Salem Village residents for their internal ill-will. The arbitrators noted that "the effects of settled prejudice and resolved animosity" had bred "uncharitable expressions and uncomely reflections tossed to and fro." Such hostility, they concluded, would "have a tendency to make such a gap as we fear . . . will let out peace and order and let in confusion and every evil work."

Then, in 1692, the local conflict took an extreme, even bizarre, turn. A witch craze swept Salem Village. At first, a few young girls accused three village women of casting evil spells on them, and soon the charges of witchcraft spread widely. Many adults joined the "afflicted" girls in making accusations against people in Salem Village, Salem Town, and surrounding communities. In part, the pattern of accusation reflected the old animosities. Of the twenty-five village residents accused of witchcraft, eighteen lived near Salem Town. Another seventeen alleged witches lived in Salem Town. Sectional conflict did not cause the witch craze, but it certainly contributed to it. Having failed to achieve their ends by using the normal forms of political discourse, Salem Village residents resorted to accusations of witchcraft. The terms of conflict in Salem were exceptional; the basic nature of the conflict, however, was quite common.

Exploitation and Expansion in the Chesapeake Region

In the Chesapeake settlements, the path of growth and development also led to conflict. By 1650, the oppressive nature of the social relations there was unmistakable. A small group of wealthy landowners exerted almost unlimited control over both the land and the laws. They relied on a largely unfree labor force to produce a single staple crop for export to England. As a result, the economy of the region expanded significantly during the second half of the seventeenth century. Tobacco exports rose from 3 million pounds in 1640, to 15 million in 1670, and then to 28 million in 1690. This remarkable increase in tobacco production demanded a terrible price in human exploitation, first of white servants and later of black slaves.

The planter elite shaped the social composition of the Chesapeake area to fit their own needs. Tobacco growers imported thousands of indentured servants from England, most of whom were young men. In the early seventeenth century, male migrants outnumbered females by a ratio of six to one; by 1700, there were still three men for every two women in the Chesapeake area. This extreme gender imbalance in the servant population reflected the high demand for field laborers. Some female migrants worked in the tobacco fields along with the men, but more often they performed domestic chores in the master's household. Few women who married raised large families because of the dangers of the environment. A typical Chesapeake woman who married in her mid-twenties would have only two or three children before she died. Moreover, because of the pressing demand for able-bodied workers, planters were not inclined to wait for a new generation to mature. They filled their needs immediately by importing more young adults. Throughout the seventeenth century, the main source of population growth in the Chesapeake region was migration rather than procreation.

The appalling conditions of life and labor in the Chesapeake area reinforced the constant need for new migrants. Up to 40 percent of servants did not survive their four- or five-year indentures. Many died from disease, others from overwork. Labor on the tobacco plantations was hard, and some masters made it even harder. According to one English observer, Virginia planters would commonly "abuse their servants there with intolerable oppression and hard usage." They used brutal methods, beating servants when they would not do enough work, driving them to do more, sometimes literally working them to death.

Such abusive treatment of workers appears surprising, even self-defeating, in a society so short on labor. The masters' behavior reflected their contempt for the lower classes of English society. Members of privileged classes on both sides of the Atlantic shared a view of indentured servants as "idle, lazy, simple people . . . such [as] have professed idleness, and will rather beg than work." One Englishman offered an even more detailed indictment:

> Among those who repair to Bristol from all parts to be transported for servants to his Majesty's plantations beyond the seas, some are husbands that have

> forsaken their wives, others wives who have abandoned their husbands; some are children and apprentices run away from their parents and masters; oftentimes unwary and credulous persons have been tempted on board by men-stealers, and many that have been pursued by hue-and-cry for robberies, burglaries, and breaking prison do hereby escape the prosecution of law and justice.

Such views confirmed the planters' belief that they were right to treat their servants harshly. In the Chesapeake area, indentured servants repeatedly found themselves tricked, mistreated, exploited, sometimes even reduced to the status of a human commodity and disposed of like any other piece of property.

Perhaps more important, planters believed that even if indentured servants were well treated, they would succumb to disease and death long before their term was up. So they worked their servants as hard as possible and then replaced them with new migrants. In a cruelly calculating sense, masters maximized the return on their investment in human capital. The early Chesapeake planters considered their servants to be not only exploitable, but expendable. An English pamphleteer warned his readers:

> All necessitated persons . . . who shall seek to shun shame or other miseries in *England,* or elsewhere, by changing Climate, and transporting themselves for this Country of *Virginia,* ought to be cautious and wary . . . that they be not abused . . . to their own destruction.

The records of the southern colonies suggest that masters did everything they could to keep their servants from tasting any of the fruits of freedom. The codes of law, like the conditions of labor, were much harsher in the Chesapeake region than they were in England. In England, servants had well-recognized customary and legal rights that constituted protection against abusive treatment and excessive exploitation. In the Chesapeake region, the planters made the laws primarily for their own benefit. The early laws of Virginia and Maryland did guarantee a servant the right to food and shelter, but did more to maintain the master's control over the servant's life. Almost all infractions committed by servants — running away, stealing, fornicating, even killing farm animals — resulted in substantially longer terms of service. One unfortunate servant who killed three of his master's hogs had six years added to his term.

Perhaps the greatest (and cruelest) irony was that if a servant lived out his or her term, the prospects for the future were reasonably promising. At the end of his indenture, a male servant was often entitled to receive "freedom dues" from his master, usually an allotment of land, tools, clothing, and food. Given this stake, a freedman could become an independent tobacco planter. Only a few former servants became wealthy planters with servants of their own, but many became reasonably successful as tobacco farmers. Even freedmen who lacked land controlled their own labor, which remained a valuable commodity. Some men became wage laborers in the tobacco fields of planters who could not afford indentured servants or who needed agricultural workers on a short-term basis. Others leased land and housing from planters in exchange for a share of the crop they produced. Even more fortunate were those freedmen who had worked as artisans, for their skills gave them the best negotiating advantage of all.

A female servant likewise had some degree of opportunity, especially given the gender imbalance in the Chesapeake region. Upon gaining her freedom, a female servant would probably be married within a few months, quite often to a man five to ten years older than herself. When he died, she would inherit at least part of his estate. As a property-owning widow, she could be more selective in choosing her next marriage partner. In general, the scarcity of able workers and available women in the early Chesapeake settlements meant that servants could look forward to the future with hope — if they lived long enough to gain their freedom.

The prospects of freedom were much more limited for another group of servants, the black slaves. As early as 1619, Dutch merchants sold a handful of Africans to Virginia planters. The exact status of these Africans remains uncertain. Perhaps they were sold as chattel slaves, but they may have enjoyed limited terms of bondage similar to those of white indentured servants. By the 1640s, however, lifelong slavery clearly existed in the Chesapeake area. A 1639 Maryland statute noted this inferior status:

> . . . all the inhabitants of this Province being Christians (Slaves excepted) Shall have and enjoy all such rights liberties immunities privileges and free customs within this Province as any natural born subject of England.

Subsequent laws and court decisions further solidified the existence of perpetual servitude that entitled masters "to have, hold, occupy, possess, and enjoy . . . Negroes forever." Increasingly, the terms "Negro" and "slave" became interchangeable.

The number of African slaves in the Chesapeake area remained rather small throughout most of the seventeenth century, especially in comparison to the number in the British West Indies. The sugar colony of Barbados, founded twenty years after Virginia, already had an African population of around 5000 by 1640; twenty years later, that number had risen to 20,000, and blacks probably outnumbered whites. In Virginia, on the other hand, slaves numbered less than 500 by 1650, roughly 3 percent of the total population of 15,000.

Plantation economics accounted for the relative insignificance of the slave population. On the average, slaves cost about twice as much as indentured servants; the price reflected the value of a person's labor for life, not just for a fixed period of years. But in the early Chesapeake settlements, life was short. Africans suffered as much as Europeans in the disease-infested environment, and they were just as likely to die soon after their arrival. For that reason, slaves were simply a bad investment: they often did not live long enough to return the master's initial outlay of capital.

After 1650, slaves became a more attractive alternative to indentured laborers. One important reason was that people in the Chesapeake region — masters and servants, whites and blacks — began to live longer. The cause of this extension of life expectancy is difficult to identify with exactness, but it generally stemmed from changes in diet and in the disease environment. As the population began to spread out into the hills and uplands, people found better food and water supplies. The decline in population concentration also led to a decrease in the chances of contagion. The Chesapeake region was still by no means a healthy place to live, but at least people could expect to survive for more than a few years. Longer life spans increased the profitability of owning slaves.

Moreover, as increasing numbers of English servants outlived their indentures, established planters cut their imports of white servants. The planters did not want to introduce further competition for themselves in the form of future tobacco growers. A decline in the birth rate in England, coupled with somewhat better opportunities for the poor there, diminished the supply of young people willing to commit themselves to bondage in the New World. Also, the discouraging reports about the treatment of servants in the Chesapeake area probably convinced many potential migrants to stay home. Changing conditions on both sides of the Atlantic thus encouraged Chesapeake planters to switch from white servants to black slaves.

The steady growth of the black population reflected the increased reliance on slave labor. In 1670, the 2000 blacks in Virginia were still a small minority among some 33,000 whites, as Chart 2.2 on p. 46 indicates. By 1700, the number of blacks had risen to 16,000, and they accounted for 28 percent of the total population. The black population of Maryland rose from 1200 to 3200 in the same period, and represented 11 percent of the colony's population in 1700.

Resentment and Rebellion

As more whites survived servitude and set up on their own, they became a threat to the planter elite. Indeed, the growing number of freedmen led to increasing competition, and eventually armed conflict, among whites in Virginia. Because wealthier, established planters occupied the best plantation sites in the tidewater region and along the main rivers, the poorer freedmen had to settle for land farther back in the interior. They were not only distant from the tobacco markets, but also more exposed to the dangers of the frontier, especially hostile Indians. Their vulnerability, both economic and physical, bred resentment against the eastern elite that controlled the economic and political power in the colony.

In 1675, the fear and frustration of the freedmen erupted in violence. Claiming they were left unprotected by the government's defense policies, frontier planters undertook their own military operations against Indians, attacking friendly and hostile tribes alike. Under the leadership of Nathaniel Bacon, an ambitious and arrogant young planter only recently arrived in Virginia, they also attacked the government itself. Bacon demanded that Virginia's governor,

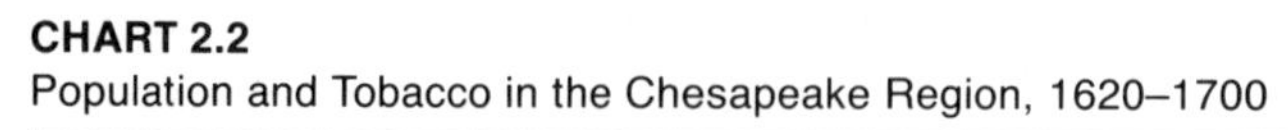

CHART 2.2
Population and Tobacco in the Chesapeake Region, 1620–1700

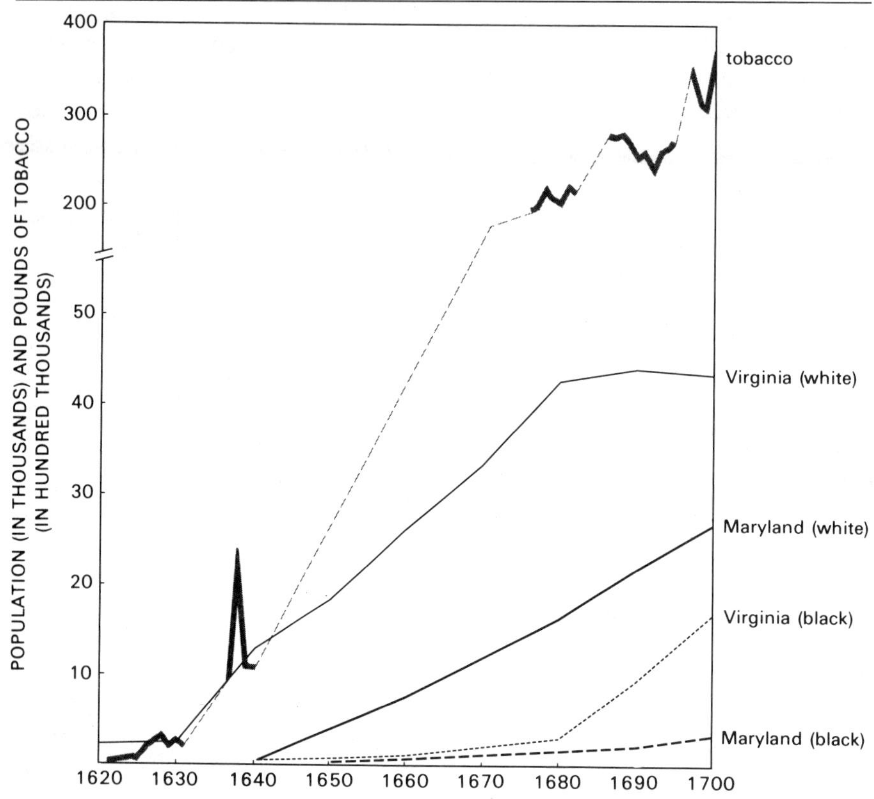

SOURCE: *Historical Statistics,* Series Z 1–19; Russell R. Menard, "The Tobacco Industry in the Chesapeake Colonies, 1617–1730: An Interpretation," *Research in Economic History* 5 (1980), pp. 157–161.

A fast-growing English market for tobacco prompted Chesapeake planters to increase production throughout the seventeenth century. After Bacon's Rebellion (1675–1676), wealthy Virginia planters were less inclined to use white indentured servants to work in their tobacco fields and purchased more black slaves.

Sir William Berkeley, grant him a commission to make war against the Indians. When Berkeley refused, Bacon led a force of 400 men on the capital at Jamestown, where the House of Burgesses was sitting. Bacon's men surrounded Berkeley and the burgesses and coerced a commission out of them at gunpoint.

Bacon's direct threat seriously undermined the governor's authority. Indeed, in the eyes of many Virginians, Bacon, not Berkeley, now controlled the colony. Bacon even put his challenge in writing. In August, 1676, he issued his "Manifesto and Declaration of the People," attacking Berkeley and his fellow "grandees" for putting their own profit from the Indian fur trade ahead of the protection of English settlers. Bacon continued his expeditions against the Indians, but now his ultimate goal was the seizure of political power. In September, he again led an armed force on Jamestown, taking the city and forcing Berkeley to flee to a boat on the James River. As long as Berkeley's ships controlled the waterways, Bacon's control of the capital meant virtually nothing. Berkeley could move freely to get supplies and reinforcements; Bacon could not. Recognizing the hollowness of his victory, Bacon abandoned Jamestown, ordering his men to burn it as they left. He retreated into the swamps to regroup, but instead his army disintegrated. Bacon came down with dysentery, and in October, 1676, he died. During the next few months, Berkeley mopped up the remains of the leaderless rebel force, killing some in battle, sending thirty others to the gallows, and driving the rest back to the frontier.

Bacon's Rebellion failed to alter the fundamental nature of the political and social order in Virginia. But the massive uprising of armed freedmen did reveal some of the underlying tensions in Virginia society. When Bacon characterized Berkeley and his followers as "grandees," he was playing upon the hostility poorer freedmen had for the planter elite. And when Berkeley's supporters wrote disdainfully of Bacon's followers as "vulgar and most ignorant people," they revealed their scorn for the lower orders. This mutual antagonism had grown throughout the seventeenth century, fed by years of strained relations between masters and servants, between wealthy planters and poorer freedmen. Bacon's Rebellion only gave this conflict its most violent expression.

Colonial Resistance to Imperial Control

Bacon's Rebellion also underscored the uncertainty of English authority. A royal governor had temporarily lost control of a colony to a band of backcountry tobacco farmers. Other English officials fared little better. The Navigation Act of 1673 had established an American customs service to enforce trade restrictions. By 1681, royally appointed customs collectors resided in every colony and were watched over by a resident surveyor. A surveyor-general of colonial customs supervised this new bureaucracy and reported American violations of the Navigation Acts to his superiors in London. The first reports were not optimistic. Rebellious inhabitants of North Carolina imprisoned customs collector Thomas Miller and tried him for treason. George Talbot, the acting governor of Maryland, picked a fight with collector Christopher Rousby and killed him. "No officer of the Customs can live in Maryland," a Royal Navy captain reported. In Massachusetts, local juries acquitted thirty-four of the thirty-six merchants accused of illegal trade by collector Edward Randolph.

The facts were clear for all to see. The Navigation Acts embodied the economic policies of the emergent English empire, but they did not reflect the political realities. Neither Chesapeake planters nor New England Puritans accepted the legitimacy of the new trade regulations. Just as the colonists had defended their settlements from Indian attacks, so they were determined to uphold their economic liberty and political autonomy.

For twenty-five years following the Restoration of 1660, royal policies also undermined English authority in the colonies. Charles II relinquished control over vast areas of his American domain. In 1663, he bestowed millions of acres of land in the Carolinas on eight proprietors, including the Duke of Albemarle, the Earl of Shaftesbury, Sir John Berkeley, and Sir George Carteret. In 1664, the king granted the newly conquered Dutch New Netherlands as well as the territory of New Jersey to his brother James, the Duke of York. Finally, in 1681, Charles gave William Penn a huge American province to serve as a refuge for Quakers, a radical Protestant religious sect.

These proprietary colonies were virtually a new form of settlement. Before 1660, joint-stock companies had settled most of the American colonies. By bringing investors (or religious sectarians) together in corporate bodies, joint-stock companies mobilized sufficient economic or spiritual resources to sustain distant settlements in the American wilderness. The only exception was Maryland, the proprietary colony granted to Lord Baltimore and his Catholic co-religionists in 1632.

The new colonies resembled the Maryland enterprise. The proprietors of the Carolinas, New York, and Pennsylvania owned all of the land; they could dispose of it to enrich themselves. Lord Baltimore and the Carolina proprietors granted some land to friends or associates and sold other plots. In addition, they issued deeds that specified payment of an annual "quit-rent," a small monetary fee that symbolized their lordship. Taken together, the quit-rents from thousands of Maryland estates provided Baltimore's descendants with a substantial income. The Penn family granted its lands in "fee simple" (with no quasi-feudal quit-rent), but prospered from the sales of its vast domain. In addition, the proprietors had the power to name the governors of their colonies and to appoint most minor officials. The only major restriction on their authority was a provision in their royal charters that all proprietary laws and regulations must "not be repugnant to the laws of England."

The paradox of increased English control over trade versus diminished royal authority over land — of the Navigation Acts versus the proprietary grants — is easily resolved. Charles II was a restored king, an exile returning to rule a people who had killed his father. The proprietorships rewarded powerful aristocrats and titled gentlemen who supported his restoration. The king's generosity to Penn likewise stemmed from a financial and political debt owed to Penn's father, Admiral Sir William Penn. In addition, Charles probably wished to rid his strife-torn kingdom of troublesome Quakers, who refused to serve in the army, pay religious taxes, or accept the authority of the legal system.

Similar political and financial calculations underlay the king's support for the Navigation Acts. Charles II pursued an expensive, even extravagant, life-style, and, like previous Stuart monarchs, refused to bargain with Parliament for additional funds. He needed the revenue produced by the import duties on tobacco and other enumerated goods to defray personal and governmental expenses. Moreover, the restrictions on foreign merchants imposed by the Navigation Acts encouraged English merchants, many of whom were Puritans, to support his Restoration regime. Although contradictory in other respects, the colonial policies of Charles II consistently supported the monarchy's best interests.

Reform and Revolt

The accession of James II to the throne in 1685 dramatically changed the character of imperial politics. James had ruled New York in an authoritarian fashion for twenty years, refusing to allow a representative assembly. English settlers in the predominantly Dutch colony composed petitions complaining that they lacked the "liberties of english men" since "a yearly Revenue is Exacted from us . . . without our consent." Only in 1683, and then with great reluctance, did James permit the formation of an assembly with limited powers.

James wished to place his subjects, both in England and America, under firm royal control. He actively supported policies for imperial reform advocated by the Lords of Trade, the effective administrative agency created in 1675 to oversee colonial affairs. To enhance royal authority in America, the Lords of Trade advocated the abolition of all proprietary charters. To uphold the Navigation Acts, they attacked the autonomy of the corporate colonies of New England. In 1684, the lords obtained a legal writ that annulled the Massachusetts Bay Charter of 1629. By 1686, similar writs extinguished the corporate charters of Connecticut and Rhode Island and the proprietary grant of New Jersey.

Then James created a vast new administrative unit, the Dominion of New England, which stretched from Massachusetts to New Jersey. To govern this territory, the king dispatched Sir Edmund Andros to Boston in December, 1686. Andros was an experienced soldier with an arrogant personality. His orders were to rule by the authority of the king's royal instructions — backed up, if necessary, by a company of English troops. The elimination of the representative General Court aroused great resentment in the Massachusetts Bay Colony. Andros's support for the Church of England likewise angered Puritan ministers and laity. Finally, the governor's threat to invalidate existing land titles and reissue them upon the payment of sizable fees (and with quit-rents now imposed) alarmed farmers and other property owners. Within a year of Andros's arrival, the Puritan colony was on the verge of rebellion.

The Glorious Revolution of 1688 in England precipitated a revolt not only in Massachusetts Bay, but in New York and Maryland as well. Even as James II imposed the authoritarian Dominion on American settlers, he pursued an equally arbitrary policy in England. The king abrogated the corporate charters of towns and guilds, imposed new taxes and levies without the consent of Parliament, and indicated that his newborn son (and heir to the throne) would be raised as a Roman Catholic. Backed by fiercely Protestant popular uprisings, leaders in Parliament forced James into exile. They invited William of Orange, a Protestant Dutch prince, and his English wife, Mary, to assume the throne as "constitutional" monarchs who would respect the rights of Parliament.

When news of William and Mary's accession reached Boston, merchants and ministers organized the Committee of Safety. The committee accused James II and his royal officials of being "bloody Devotees of Rome" and issued a Declaration of Grievances. Overpowering Andros's troops, they seized the governor and sent him back to England. Subsequently, an enraged mob forced the jailing of Joseph Dudley, one of Andros's prominent local supporters. Capitalizing on this fluid political situation, non-Puritan property owners successfully demanded the right to vote from the Committee of Safety. A new charter, granted by King William III in 1692, guaranteed the franchise to property owners, without regard to membership in the Puritan church. The charter also transformed Massachusetts into a royal colony controlled by a crown-appointed governor and a representative assembly.

In New York, the revolt against the Dominion of New England sparked a violent ethnic confrontation. Led by Captain Jacob Leisler, a minor merchant of German origin, the New York City militia deposed Francis Nicholson, the lieutenant governor of the Dominion. Staunchly Protestant Dutch artisans joined English Puritan settlers on Long Island in attacking Dominion officials as "Popish Dogs and Devils." Then the Dutch residents reclaimed some of their former political power. They supported the interim provincial government headed by Leisler and hailed his release of debtors from prison. To protect their economic interests, Dutch (and some English) workers elected six artisans to the board of aldermen. Some city residents resorted to mob violence against English merchants and their Dutch partners. They "take peoples' goods out of their houses," claimed one merchant, "and if hindered by the Justices of the peace, they come with great numbers and fetch it out of the Justices house by force."

Merchants blamed Leisler for this upsurge of ethnic and economic antagonism, accusing him of deluding "poor ignorant innocent and senseless people." Indeed, some wealthy men labeled the militia captain a "Masaniello," after a fishmonger who led a bloody revolt in Naples, Italy, in 1647. And the English merchants took their revenge when a new royal governor arrived in New York in 1691. An all-English jury convicted Leisler and his brother-in-law Jacob Milbourne of treason and ordered them to be hanged and then decapitated. These executions signified the permanence of English rule in New York.

The revolt in Maryland likewise revealed profound economic and religious tensions. On James II's downfall, Protestant settlers established "An Association in Arms for the Defense of Protestant Religion." Led by John Coode, a small-scale planter who proudly called himself a "Masaniello," this association established a new government. It expelled the Catholic appointees of Lord Baltimore from public office and prohibited Catholics from conducting public religious rituals. The rebels accepted a royal governor and, in 1715, a proprietary regime headed by a Protestant descendant of Lord Baltimore.

For all their diversity, the American revolts of 1689 had three common characteristics. First, they were fiercely anti-Catholic, underlining the intensely

Protestant values of Massachusetts Puritans, New York Dutch, and most Chesapeake Bay planters. Second, these uprisings revealed widespread popular discontent. Dutch artisans, non-Puritan merchants and farmers, and Maryland yeomen took the opportunity offered by the Glorious Revolution to protest against the oligarchic and authoritarian character of the various colonial regimes and to demand political concessions. Finally, these revolts expressed a common antipathy to English control, in the form of either the Dominion of New England or the authoritarian proprietorships.

Viewed from this perspective, the revolts embody many of the primary themes in the history of England and America in the seventeenth century. They underline the importance of religion in the colonization movement. Like the Puritan Revolution and the Glorious Revolution in England, the uprisings also revealed the tenuousness of political authority in an era of rapid social change. Absolutist kings, religious magistrates, or aristocratic oligarchies headed both societies, but their legitimacy was open to challenge. Traditional authority was particularly vulnerable in the American wilderness, where nearly the entire population was armed with guns — to guard cattle from wolves, to protect cornfields from wandering horses and pigs, to overawe and kill Indians.

Indeed, with this resort to violence, English and American history came together in a single story. Beginning in 1607, adventurous and determined English men and women invaded Indian America. By 1676, the colonists had vanquished most of the seaboard tribes. Then, almost immediately, they resisted the imposition of external economic and political controls. By 1700, the English colonists in America had established prosperous and semiautonomous societies. Prosperity and autonomy would be two major themes that shaped the history of the American colonies throughout the eighteenth century.

3

The American Colonies in the Atlantic Economy

Even though the English government failed to establish firm political control over the American colonies, England still derived enormous benefit from them. These far-flung outposts of empire made a critical contribution to England's expanding role in the Atlantic trade system. Beginning in the earliest stages of settlement, the colonies provided valuable raw materials and foodstuffs that could not be produced in England. The export of American staples, such as sugar and tobacco, fed a growing consumer demand in England; the resale of these goods to Europe also created additional wealth for England. Moreover, the colonies became important markets for English exports, especially manufactured and processed goods, such as tools, agricultural implements, and cloth. The growth of the colonial economy also generated demand for luxury items such as tea, wine, and silk.

England's benefits from its colonial trade went far beyond the simple exchange of goods. Investors who provided capital for colonial enterprises, especially sugar plantations in the West Indies, received a substantial return. By the time of the American Revolution, English absentee owners owned sugar plantations valued at £15 million; the income from these investments was more than £1 million a year. English merchants also profited handsomely from the colonial trade by providing the shipping and financial services to American producers. During the period from 1670 to 1730, the net returns on capital invested in the rapidly expanding shipping industry averaged between 5 and 10 percent per year. Finally, the importation of raw materials and unprocessed foodstuffs from the colonies led to the creation of new enterprises in England. Sugar refining, tobacco processing, and (later) cotton weaving were all lucrative businesses based on colonial crops. By employing thousands, these businesses increased the size of the domestic market for manufactured goods. They also transformed imported commodities into more valuable products, for sale both to Europe and back to the American colonies.

The American colonies also benefited from this relationship, but in different ways and to different degrees. Especially in the first century of settlement, economic growth depended on staple exports. Those regions that produced the most valuable export crops — the sugar islands of the West Indies, the lowland rice country of South Carolina and Georgia, and the tobacco-rich Chesapeake area — experienced the greatest immediate impact of transatlantic trade. Each of these regions developed a one-crop economy — a monoculture — to meet the high market demand. In some of the southern colonies, the concentration on one staple product inhibited economic diversification; continued profit came at the cost of economic underdevelopment and dependency. The northern colonies, by comparison, had no single dominant export commodity. Instead, they produced a variety of goods. More important, they carried on their own mercantile and manufacturing activities. By the middle of the eighteenth century, the northern colonies, although not economically self-sufficient, were much more fully developed than were their southern counterparts.

There was no single colonial economy. Rather there was the development of several regional economies. These economies were distinct from each other, defined by different combinations of environmental factors, natural resources, and human settlement patterns. Yet they were also part of an interrelated imperial system, connected not only to England but increasingly to each other as well.

The West Indies: Sugar and Slaves

No mainland region or colony played as important a role in the British imperial system as did the sugar islands of the West Indies. Sugar, like tobacco, was a relatively new product in Europe. First a luxury item, then seemingly a necessity, it became a critical commodity in the transatlantic trade. Despite price fluctuations that gradually followed a downward course, the sugar market remained one of the most profitable sectors of the colonial economy throughout the seventeenth and eighteenth centuries. By filling the high demand for sugar and sugar products (especially molasses and rum), West Indian planters made a major contribution to England's national wealth, as well as massive fortunes for themselves.

The English were comparative latecomers to the West Indies. Beginning in the second half of the sixteenth century, English explorers and privateers sailed the Caribbean, which was still essentially a Spanish domain. In 1624, with the establishment of a foothold on St. Catherine's (later St. Kitt's), the English began to make permanent settlements on the islands. During the next decade, they took control of Barbados (1627), Nevis (1628), and Montserrat and Antigua (1632). Finally, the capture of Jamaica from the Spanish in 1655 established the English as a major presence in the West Indies.

The islands provided not only valuable bases for naval operations against Spanish shipping, but also excellent territory for agricultural development. The warm, wet climate of the islands meant a year-round growing season and an environment in which plant life flourished. In the early years of settlement, English planters concentrated on growing food for their own consumption and tobacco and cotton as cash crops. By the 1640s, however, they had turned to sugar production. From that point on, the English West Indies experienced a sugar boom that rivaled and soon surpassed the tobacco boom of the Chesapeake region.

Barbados, one of the larger islands (166 square miles), with 10,000 acres of arable land, quickly became the center of the sugar economy. "There is not a foot of land in Barbados that is not employed even to the very seaside," reported one observer. According to another, it was "the most flourishing Island in all those American parts, and I verily believe in all the world for the producing of sugar . . . and their sugar works are brought to that maturity that each work for the space of eight months in the year can and doth yield 1500 lb. of sugar each 24 hours." By the 1650s, Barbadian planters were exporting 3000 to 7000 tons of sugar annually; later in the century, sugar production sometimes exceeded 10,000 and even 15,000 tons a year. With the expansion of production, the London price of sugar declined gradually, from a high of 100 shillings per hundredweight (112 pounds) to a low of less than 20 shillings. After the 1680s, however, the price generally remained between 20 and 40 shillings per hundredweight throughout the eighteenth century, as Table 3.1 on p. 56 shows.

Whatever its price, the profit on sugar was good enough to encourage island planters to devote almost all their land to growing cane. "Men are so intent on planting sugar," a West Indian told Governor John Winthrop of Massachusetts Bay in 1647, "that they had rather buy food from you at very dear rates than produce it by labor, so infinite is the profit from sugar." The most successful sugar planters became remarkably wealthy, easily as rich as the wealthier landed aristocrats in England. As late as 1776, Adam Smith noted in *The Wealth of Nations:* "The profits of a sugar plantation in any of our West Indian colonies are generally much greater than those of any other cultivation that is known either in Europe or America."

Before sugar planters could savor their large profits, they had to incur considerable costs. Turning raw cane into a transportable form such as molasses, rum, or muscovado (a dark, semirefined granular sugar) required an initial investment in processing equipment — a mill, boiling pots, and sometimes a distillery. Moreover, sugar planters needed a handful of skilled workers to operate this equipment and a much larger work force to labor in the cane fields.

From the earliest days of the sugar industry, most of the field workers were African slaves. Estimates of slave imports suggest that between 1640 and 1700

TABLE 3.1
English Sugar Imports (Annual Averages, Selected Years, 1698–1757)

Years	Total Imports (Cwt.)[a]	Domestic Consumption (Cwt.)[a]	Percentage Reexported	Shillings Per Cwt.[b]
1698–1700	471,000	295,000	37.5%	41s
1716–1720	653,000	493,000	24.5	33s
1728–1732	926,000	797,000	14.0	
1733–1737	806,000	772,000	4.2	18s
1748–1752	896,000	857,000	4.4	
1753–1757	1,092,000	1,043,000	4.5	37s

SOURCE: Richard B. Sheridan, "The Molasses Act and the Market Strategy of the British Sugar Planters," *JEH* 17 (1957).
[a]Short hundred weight = 100 lbs.
[b]Long hundred weight = 112 lbs.

West Indian sugar planters grew wealthy as a result of the high demand for sugar in England. English imports doubled between 1700 and 1750, and domestic consumption rose even more dramatically because (as Table 3.1 shows) English sugar exports to Europe declined markedly.

over a quarter of a million Africans were landed in the English West Indies — about half of them in Barbados (134,500) and the other half divided between Jamaica (85,100) and the Leeward Islands (44,000). In the eighteenth century, slave importation continued to increase, as shown in Table 3.2. The number of slaves imported annually reached 10,000 in 1740 and rose to 13,000 by 1770.

This massive slave importation came about to offset the high death rate in the West Indies. African slaves, like the English inhabitants, fell prey to a variety of infectious diseases. Some of these maladies — hookworm, leprosy, elephantiasis, and encephalitis — the Africans brought with them on the slave ships. Indeed, over 10 percent of the slaves shipped by the Royal African Company between 1680 and 1734 died before reaching the islands. Those slaves who survived the ocean passage were still likely to be infected through contact with their fellow slaves or with whites, the main carriers of tuberculosis and venereal disease. The Caribbean environment itself held the dangers of yellow fever, and the filthy conditions of plantation quarters took a similar toll of the enslaved population.

The human threat to slaves was as great as the environmental one. Like Virginian tobacco planters, West Indian masters sought to get the most work out of their labor force in the shortest possible time. The year-round growing season on the sugar plantations gave slaves little rest from their work routines. The burden of the labor and the brutality of the masters killed many slaves. Even more tragically, other Africans killed themselves to escape the inhuman conditions of plantation life. "I have lost in one year a dozen new Negroes by dirt eating though I fed them well," reported one Jamaican planter. "When I remonstrated with them, they constantly told me, that they preferred dying to

TABLE 3.2
The Black Population of the British West Indies

Years	1703–1712		1720–1734		1745–1748		1756–1762		1773–1778
Black population	122,000		158,300		218,900		286,800		335,277
Numerical increase		36,000		60,600		67,900		48,400	
Importation during interval		151,300		189,900		245,400		266,800	
Excess of imports over numerical increase		115,300		129,300		177,500		218,400	

SOURCE: Compiled from Philip Curtin, *The Atlantic Slave Trade* (University of Wisconsin Press, 1969) Tables 39 and 40; and Frank W. Pitman, *The Development of the British West Indies, 1700–1763,* (Shoe String, 1967), pp. 372–383.

Deadly epidemic diseases and brutal working conditions on the West Indian sugar plantations combined to produce an extremely high death rate among slaves. Even though white planters imported more than 830,000 Africans between 1703 and 1778, the black population increased by only 210,000 during that time.

Sugar Production in the West Indies. This engraving shows slave laborers on an eighteenth-century plantation using simple technology to process raw sugar cane for export. Slaves had to labor strenuously to plant and harvest the cane and were supervised closely by the plantation master during the sugar-making process. (*Source: The Granger Collection*)

living." "A new Negro is not very tractable," another white observer commented. "He languishes for some time, then dies, and his death is ascribed to the climate, which has been but little to blame." Throughout the seventeenth and eighteenth centuries, one of a sugar planter's greatest capital costs was the replacement of dead slaves.

The amount of capital required for land, labor, and equipment had important consequences for the white population of the West Indies. Those planters who could afford the initial investment and absorb the replacement costs had a great advantage over their poorer neighbors. The more successful planters soon gained a remarkable degree of control over the island colonies, especially Barbados. In 1680, the leading 175 planters on Barbados constituted only 7 percent of the property holders, yet they controlled well over half of the island's wealth. On the average, members of this elite each had 210 acres of land and 115 slaves. They also held virtually all the major political and military offices. Below them were several thousand poorer whites, including more than 1000 property holders who each owned fewer than 10 acres and no slaves.

Many members of this planter elite eventually chose to leave the island, returning to England to be absentee owners. Their continued control of the island's resources limited the economic opportunities of the poorer whites and forced many of them to leave as well. Estimates vary widely, but some 10,000 to 30,000 whites departed from Barbados in the second half of the seventeenth century, usually headed for the smaller, less developed islands of the West Indies, the Carolinas, or even back to England. The white population of Barbados dropped significantly, from 30,000 in 1650 to 13,000 in 1710.

The decline of the white population was the final step in the formation of a slave society on Barbados. By 1700, the 50,000 slaves on Barbados represented three-fourths of the island's population. The other island colonies followed the same pattern, with even more extreme results. In 1750, the ratio of blacks to whites on Barbados was four to one; on the Leeward Islands, it was seven to one; and on Jamaica, ten to one. The sugar islands were the first major English possessions in the New World to have a sizable black majority. They would not be the last.

The Lower South: The Rise of Rice Culture

The economic and demographic development of Barbados had an effect that extended far beyond the sugar islands. As early as the 1660s, uprooted Barbadians began to colonize the Carolinas with the encouragement of the proprietors. In the early 1670s, Barbadian emigrants and their slaves comprised almost half the inhabitants of the first South Carolina settlements. Well into the next century, the new southern colony clearly bore the stamp of the West Indies, both culturally and economically. It was, as one historian has put it, "the colony of a colony." South Carolina did not produce sugar, but it did reproduce a West Indian society. Like its parent colony, it produced a staple crop using the labor of an enslaved black majority and was dominated by a white planter elite.

From the very beginning of settlement, white South Carolinians searched for a suitable export crop. They first turned to the coastal woodlands, where timber and furs provided valuable commodities for trade. Wood products — pitch, tar, resin, turpentine, and ships' masts — were especially important to the English navy, and naval stores remained an important part of South Carolina's export trade up to the time of the American Revolution. At the urging of the proprietors, seventeenth-century South Carolinians also produced sugarcane, tobacco, and cotton, as well as more exotic products like ginger, grapes, and olives. Their first major success was cattle raising. The abundance of forage land allowed cattle to be grazed widely, and the herds reproduced and expanded at a remarkable rate.

Unfortunately, the agrarian background of most English settlers provided them with almost no experience as cattle graziers. Many of their slaves, however, did have such experience. Due largely to the skills of African-born horsemen and herders, cattle raising became the leading enterprise in seventeenth-century South Carolina; it enabled the colonists to develop a sizable trade with West Indian planters, who had a constant need to import food. In return, South Carolinians imported increasing numbers of slaves.

Cattle raising was only an intermediate step in the development of the

South Carolina economy. The capital and labor acquired in the cattle trade provided the foundation for the production of a much more important staple crop, rice. The climate and topography of South Carolina, especially in the coastal lowlands and swamps, offered favorable conditions for growing rice. Early planters in the colony experimented on a small scale with rice, but never produced enough to make it a valuable export commodity. As with cattle herding, the English settlers had little experience in rice growing. Here again, blacks made a critical contribution to white success. Slaves from the west coast of Africa had long been familiar with rice planting; they provided not only the physical labor but also the agricultural knowledge that made South Carolina a major rice-producing region. The cruel irony was that the expansion of rice production led to the importation of more and more slaves.

Both rice production and the slave population expanded dramatically during the eighteenth century. Blacks had always constituted between one-fourth and one-third of South Carolina's population; as early as 1708, however, their numbers had surpassed those of whites. By the 1730s, blacks accounted for over 60 percent of the population, and they remained a clear majority throughout the eighteenth century. As one European migrant commented in 1737, "Carolina looks more like a Negro country than like a country settled by white people." In comparison, slaves were still a minority of the populations of Virginia and Maryland by the middle of the eighteenth century, comprising just over 40 percent and 30 percent respectively. The concentration of the slave population in South Carolina's rice-growing districts was even more striking. In St. George's parish, for instance, the number of slaves per white household rose from eight in 1720, to twenty-four in 1741. In some parts of the rice belt, blacks outnumbered whites by nine to one.

The exploitation of slave labor fed the expansion of the rice economy. Plantation owners used slave labor to construct and care for an extensive system of dikes and irrigation ditches that greatly expanded the acreage devoted to rice production. South Carolina's rice exports increased more than tenfold between 1710 and 1730, rising from 1.6 million pounds to 18 million pounds. These exports continued to rise, although not as sharply, until the American Revolution. The export of Carolina rice to southern Europe kept the demand, and therefore the price, high throughout the eighteenth century. Indeed, by 1770, the price of rice had risen 10 percent over its 1730 level.

Planters in South Carolina increased their wealth even more by producing indigo, a highly valued dye for cloth. Since 1650, English manufacturers had imported this bluish-purple dye from the West Indies. Then, in the 1740s, a young woman named Eliza Lucas, while temporarily in charge of three of her father's plantations, introduced large-scale indigo cultivation to the southern colonies. Indigo was an excellent complement to rice as a staple crop because it grew on high lands unsuitable for irrigation and rice production. Even more importantly, the planting and harvesting of indigo fell at slack times in the rice-growing season. South Carolina planters were thus able to utilize both land and labor more efficiently. The creation of a year-round work routine reduced the opportunity for idleness — and perhaps rebelliousness — among slaves. It also increased the master's return on investment in slave labor. Eighteenth-century observers estimated that slaves on the rice and indigo plantations paid for themselves in two to three years.

The South Carolina planter elite benefited enormously from slavery and colonial trade. By 1750, rice and indigo were, respectively, the third and fifth most valuable commodities exported from the mainland colonies. The expansion of the export trade made Charleston one of the busiest and wealthiest of American port cities. Many members of the gentry spent at least part of the year there, exchanging the social isolation and health risks of plantation life for the more convivial and invigorating environment of the city. They also spent a good deal of their money there, buying the latest and best imported goods. Indeed, the rise in English imports into the lower South as a whole, as shown in Chart 3.1, was greatly stimulated by the consumer demand of the small but extremely wealthy planter elite.

The South Carolina gentry held immense power. In expanding their control of land and labor, they drove small, independent white farmers to the fringes of the rice district. Throughout the colonial era, no other mainland col-

CHART 3.1
Population and Imports in the Carolinas and Georgia, 1700–1780

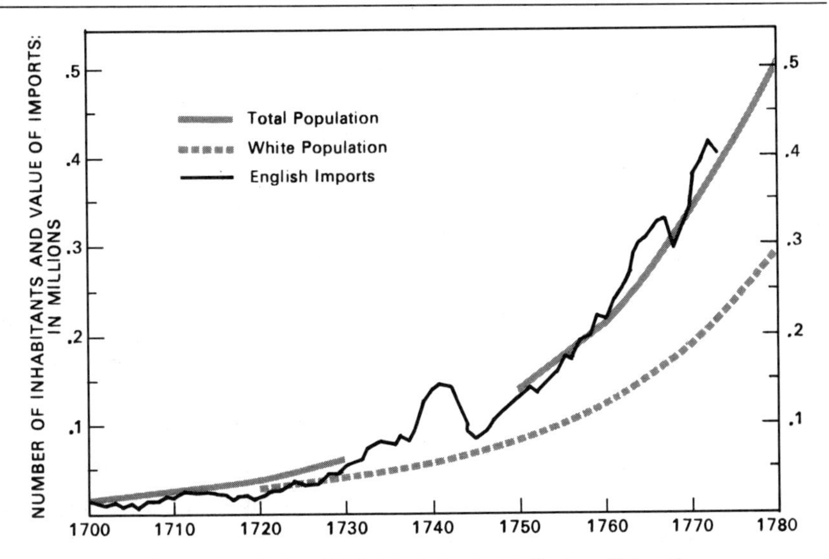

SOURCE: *Historical Statistics,* Series Z 15–17; John J. McCusker, "The Current Value of English Exports," *WMQ* 28 (1971), Table III.

Each of these three colonies had a distinct demographic character. In North Carolina, there were far more whites than blacks; in Georgia, the two races were roughly equal in numbers by 1770; in South Carolina, whites were a minority. In all three colonies, however, whites consumed almost all of the goods imported from England.

ony had the extreme racial and economic imbalance that was characteristic of South Carolina.

Georgia mirrored South Carolina economically and socially, but it never reached the same level of prominence in the colonial era. Georgia was established comparatively late, in 1733. Partly intended to provide a southern line of military defense against the Spanish in Florida, the new colony was also an enlightened social experiment. Seeking to create a refuge for England's poor, James Oglethorpe and a group of philanthropic trustees successfully petitioned the king for a charter. They envisioned a colony of small, self-sufficient farms inhabited by independent landowners and indentured servants. Accordingly, the trustees limited most individual land grants to 500 acres and outlawed slavery.

Within a decade, however, a significant number of Georgia settlers, including many South Carolinians who owned Georgia land, campaigned for the introduction of slavery. Like South Carolina, Georgia had extensive coastal lowlands. Because of the rigors of rice cultivation and the dangers of the lowland environment, most whites were unwilling to labor in the rice fields. They wanted slaves to do the work. The proslavery faction in Georgia eventually had its way. In 1750, the trustees eliminated the prohibition against slavery, and the restriction on the size of individual land grants fell soon afterward.

After 1752, when the trustees surrendered their charter and Georgia came under direct royal control, the colony became increasingly a racially and economically divided plantation society. The vast majority of whites carried on small-scale, semisubsistence farming in the backcountry, while a few wealthy slaveowners controlled rice production along the coast. In 1750, one-fifth of Georgia's population of 5200 was black; by the early 1770s, however, the more than 15,000 blacks almost equaled the white population. In the rice-growing areas along the coastal lowlands and the inland rivers, blacks constituted a significant majority. Rice accounted for about a third of the total value of exports between 1752 and 1776. The export trade as a whole grew steadily during that period. The number of ships clearing Savannah increased threefold, and the value of exports reached £121,000 in 1773. As Table 3.3 indicates, Georgia contributed to a steadily expanding rice trade in the eighteenth century.

North Carolina likewise participated in the export economy, but with much more difficulty and with much less profit than the rice colonies. Settlers there tried to find a suitable staple crop for the Atlantic market, but they were largely unsuccessful. They exported a variety of foodstuffs — peas, Indian corn, meat, and dairy products — but not enough to constitute a significant source of income to the colony. For the most part, North Carolina remained a colony of small farmers who produced primarily for their own consumption or for a regional market.

Some wealthier lowland planters in North Carolina used slave labor to grow the kinds of cash crops that were common elsewhere in the South, especially tobacco, rice, and indigo. Hanover and Brunswick counties, both in the Cape Fear region, had black majorities. Elsewhere along the coastal lowlands,

TABLE 3.3
Rice Production in Georgia and South Carolina

Year	Millions of Pounds Exported[a]	Price Per Pound[b]	Market Value	Population	Rice Exports Per Capita
1730	16.8	1.07 pence	£ 74,400	30,000	50 shillings
1740	30.6	1.07 "	136,400	47,000	58 "
1750	30.3	1.13 "	142,600	69,200	41 "
1760	41.9	1.04 "	181,500	103,500	35 "
1770	76.9	1.17 "	374,900	147,500	51 "

SOURCE: *Historical Statistics*, Ser. Z 262–266 and 340.
[a]Five-year averages.
[b]Philadelphia prices.

In the lower South, rice production and population increased almost fivefold between 1730 and 1770. Since the market price for rice also rose, the white planters of the region enjoyed great prosperity up to the time of the American Revolution.

slaves made up less than one-third of the population. Tobacco exports rose from around 100,000 pounds in 1753 to 1,605,000 pounds in 1772, but tobacco farming was generally confined to the northeastern corner of the colony near the Virginia border. In all, fewer than 8000 acres were devoted to the production of staple crops.

In the late colonial era, the main export commodities were forest products, primarily sawed timber and naval stores. In fact, in 1768, North Carolina accounted for 60 percent of the naval stores sent from the North American colonies to England. However, North Carolina's forest products were low-quality, low-profit commodities. Between 1768 and 1772, the annual value of North Carolina's exports amounted to an estimated £1.17 per white inhabitant. By comparison, Virginia and South Carolina, which dealt in much more profitable goods, had annual per capita export values of £2.82 and £9.13, respectively.

Moreover, about half of North Carolina's small volume of exports went first to Virginia or South Carolina before being shipped overseas. The narrow sandbars and barrier islands that stretched for several hundred miles along the North Carolina coast formed an effective barrier to shipping, and the colony had no easily accessible port like Charleston. "Every Circumstance that places us on a more unfavorable footing than our neighbors is to enrich them at our expense," wrote Governor Josiah Martin in 1773. "We need but to look to the Colonies next adjoining us, to discover the invidious distinction."

The Chesapeake Region: Tobacco and Transition

The planters of the Chesapeake region found their cash crop early in the seventeenth century, and they remained largely committed to it throughout the eighteenth. The remarkable expansion of the seventeenth-century tobacco in-

TABLE 3.4
Virginia, Maryland, and Delaware Tobacco Exports, 1700–1770

Year	Millions of Pounds Exported[a]	Price Per Pound[a]	Market Value	Population	Tobacco Exports Per Capita
1700	32	2.3d	£306,000	90,600	67s 7d
1710	30.4	—	—	124,700	—
1720	35.6	1.02	151,300	159,200	19s 0d
1730	43.8	1.1	200,700	214,300	18s 7d
1740	52	1.2	255,800	316,400	16s 2d
1750	66.2	1.4	330,700	400,800	16s 7d
1760	69.8	1.5	436,250	535,200	16s 5d
1770	83.8	2.0	700,000	685,100	20s 3d

SOURCE: *Historical Statistics,* Ser. Z 12–14, 230–237, 339.
[a]Five-year averages.

Tobacco prices remained steady at one pence per pound between 1660 and 1740 (except during wartime, as in 1700). They then rose quickly, bringing prosperity to the Chesapeake region in the decades preceding the American Revolution.

dustry slowed in the last quarter of the century, and tobacco production remained fairly stable for about forty years. In the 1720s, the tobacco economy once again began a long period of generally steady growth, as Table 3.4 shows; British imports increased about 2 percent per year up to the time of the American Revolution.

The main cause for this expansion came from increased European demand — or more accurately, from increased British penetration of the European market. Resale of American tobacco out of Great Britain rose from 50 percent of the 15 million pounds shipped from the colonies in 1670 to 85 percent of the 100 million pounds sent a century later. British domestic consumption of tobacco doubled during this period, but foreign demand increased tenfold. Working through Scottish merchant houses, the French government began to import Chesapeake tobacco in the 1720s, and within twenty years France had surpassed Holland as the prime recipient of British reexports.

This sustained increase in international demand stimulated the entire economy of the Chesapeake region. The districts along the York and Rappahannock rivers in Virginia, the traditional centers of the tobacco industry, increased their production from 18 to 30 million pounds between 1714 and 1774; their share of total output, however, dropped during that period from 59 percent to 31 percent. The Piedmont area and the valleys of the Potomac and James Rivers took their places as the leading tobacco regions. By the 1770s, the shipments just from the customs districts on the James River amounted to 52 percent of the total tobacco output.

An elaborate international framework of credit operated by Scottish merchant houses supported this expansion into the interior. Confident that there

Early British Advertisement for Virginia Tobacco. This trade card from about 1750 is typical of brand advertising of the era. The great expansion in the production of colonial tobacco forced British merchants to stimulate demand in order to absorb the ever-increasing supply. (*Source: George Arents Collection, New York Public Library.*)

would be a continuing European market for low-priced Virginia tobacco, Scottish merchants were only too willing to provide the capital resources needed to increase output. Beginning in the 1740s, they established dozens of subsidiary stores and shipping outlets in newly settled areas. The agents who managed these establishments granted credit to planters for their purchases and took tobacco in payment. Both wealthy planting families and poor white farmers relied increasingly on the availability of Scottish credit. By the time of the American Revolution, thousands of Virginians had gone deeply into debt.

Penetration into the European market did not guarantee wealth for everyone. Competition from the Mediterranean kept tobacco prices low in European markets. Moreover, the competition among thousands of Chesapeake growers made it virtually impossible for them to act cooperatively to control production in order to raise prices. Although the economy of the Chesapeake as a whole benefited from the expansion of production, many individual planters did not. Even tobacco planters with large estates were unable to accumulate fortunes to rival those of the rice growers in South Carolina or the sugar growers in the West Indies.

Declining prices and deteriorating lands forced some tobacco growers to look for alternative crops. As early as the 1680s, planters on the lower eastern shore of Maryland and in southeastern Virginia had begun to produce Indian corn and wheat for export. The settlement of the Shenandoah Valley in the eighteenth century added another grain-producing region. By the 1770s, Virginia was exporting 400,000 bushels of wheat per year, one-fourth as much as the total produced by the breadbasket colonies of Pennsylvania and New York. Grain and wood products accounted for about one-quarter of the Chesapeake region's export commodities.

The expansion of tobacco and grain production contributed to the development of an increasingly successful export economy in the Chesapeake area. Between 1700 and 1770, when the population of the region grew at a remarkable rate from 90,000 to 685,000, per capita exports remained steady, as shown in Chart 3.2. But aggregate figures alone do not measure the true significance of the trade. In the years after 1725, the value of commodity exports grew at a far faster rate than did the *white* population — the segment that would, after all, benefit most from the increased earnings. Diversified crop production helped keep the Chesapeake region in a position of prominence in the Atlantic market.

Growth versus Development

The value of exports provides only a one-dimensional picture of the nature of the economy of the southern colonies. Staple production brought considerable wealth to whites, and it supported the geographical and numerical expansion of the population throughout the colonial era. But economic *growth* did not bring about extensive *development*. The southern colonies did not achieve the diversity of interrelated enterprises that identify a mature and generally self-sustaining economy. The continued and almost single-minded concentration on increasing agricultural exports inhibited innovation and expansion in other economic sectors, especially commerce, finance, and manufacturing. By 1750, the southern colonies remained essentially what they had been a century before — dependent on more highly developed economies for capital, trade and managerial services, technology, manufactured goods, and in some cases even food.

The slowness of urbanization was the clearest indicator of the limits of southern development. Only Charleston, South Carolina, with a population of 12,000 in 1775, became a major city during the colonial era. Baltimore, Maryland, and Norfolk, Virginia, grew from small settlements of a few hundred inhabitants in 1750 to sizable port towns of approximately 6000 inhabitants each in 1775. The other cities of the region — Annapolis, Maryland; Williamsburg, Virginia; and Savannah, Georgia — remained little more than large towns that played relatively minor roles in the economic life of their respective colonies. In the Chesapeake area, where the system of rivers provided numerous outlets to the Atlantic, tobacco producers had little need for a single port city. They traded their tobacco directly with merchants who supplied them with goods from London, Glasgow, and other British cities. Even Charleston was little more than a collecting and shipping point for export goods. Planters brought their rice and indigo to the port, where these plantation products quickly moved into English

CHART 3.2
Population and Imports in the Chesapeake Region, 1700–1780

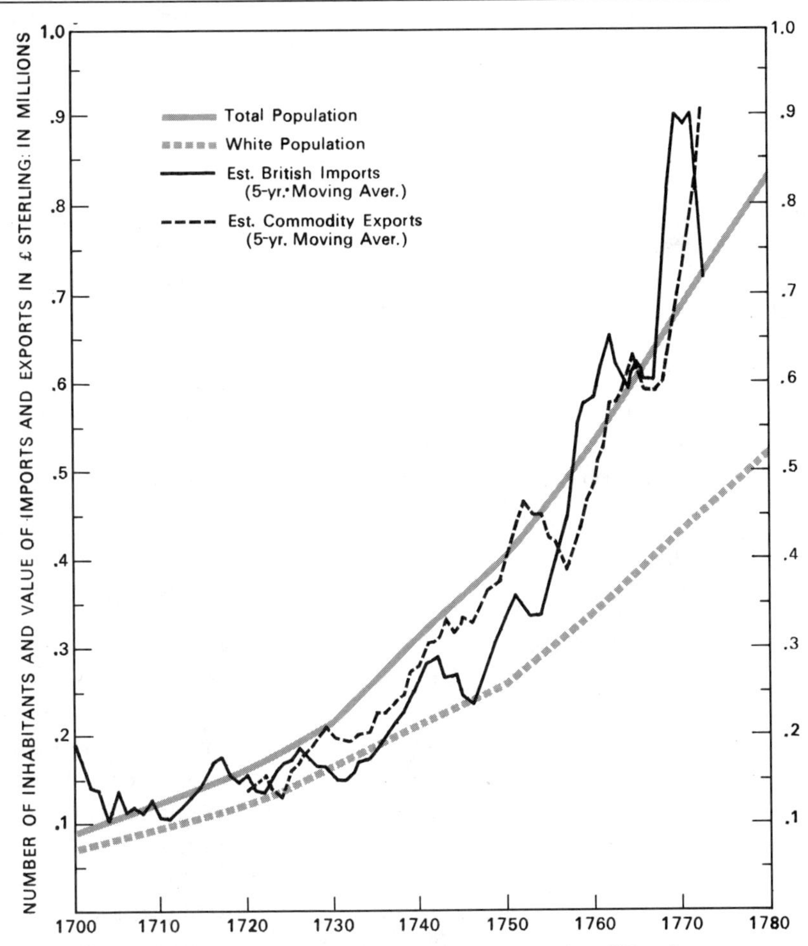

SOURCE: *Historical Statistics*, Series Z 12–14; John J. McCusker, "The Current Value of English Exports," *WMQ* 28 (1971), Table III.

In the Chesapeake region, the export of tobacco (and, after 1740, of wheat and corn) paid for imported cloth, furniture, tools, and utensils. Per capita consumption of British goods rose steadily during the pre-Revolutionary decades, especially among the wealthier slave-owning families.

hands. British merchants provided both the shipping and marketing services, and South Carolinians never gained control of the export business. Most leading planters were content to purchase British manufactured goods and luxury items and then return to the comfort of their plantations and townhouses.

Charleston, like other southern cities, had a number of artisans and shopkeepers who supplied some goods and services to the urban population. In the early part of the eighteenth century, in fact, slaves created a kind of black-market economy in handcrafted goods and garden vegetables. The animosity and apprehension of whites eventually curtailed even these low-level enterprises, however, and blacks were generally denied the opportunity to participate directly in the economy.

The lack of well-established urban centers, the control of merchant activity by British capitalists, and the subjugation of a large slave population that had little consumer power were not conditions conducive to a process of economic diversification and development. More important, the control of the region's wealth by a small planter elite ultimately had a stultifying effect on the economy. Southern planters concentrated their energies on the management of their own estates and the creation of new ones for their offspring. Rather than reinvesting their considerable profits in new technologies and economic endeavors, they plowed their money back into the original sources of their wealth, land and slaves. The importance of the choices they made in the eighteenth century would not become fully evident until the nineteenth. Even in the colonial era, however, the growth of the agricultural economy of the South stood in sharp contrast to the development of commercial capitalism in the North.

New England: Merchants and Markets

In the first years of settlement, New England offered little prospect of rapid economic development. For one thing, the Puritan commitment to communalism emphasized social equality over economic expansion. By dividing the land into a number of individual family holdings, early New England townspeople created a society composed of small-scale, self-sufficient producers. Moreover, New England's thin, rocky soil and long, harsh winters made farming a difficult, sometimes discouraging, enterprise. Most farm families had enough to do to provide for themselves, much less produce a surplus for trade. Ironically, the eventual success of the New England economy resulted in large part from the initial limitations on agricultural production. Because the early inhabitants did not grow a staple crop as valuable as sugar, tobacco, or rice, they were forced to discover diverse ways to create wealth.

Throughout the seventeenth century, New England's exports never equaled its imports. The growing population continually relied on goods purchased from England. In order to finance the trade deficit, New Englanders had to pay for their purchases partly with cash — a very scarce commodity. Most of the money settlers brought with them to New England soon went back on English ships and wound up in the pockets of English merchants.

The expansion of trade between England and New England did lay the foundation for future economic development. Taking advantage of whatever opportunities they could find — in cooperative ventures with English merchants or trade with other colonies — New England merchants gradually made inroads on the dominance the English had over colonial trade. These local merchants increasingly provided the goods, the ships, and the entrepreneurial energy that would make New England, and especially Boston, the first commercial center of the American colonies.

The role of New England merchants in the fishing industry exemplifies this process. Beginning in the 1540s, the waters off the coasts of Newfoundland and New England were a prime spot for European fishing fleets. In the 1620s, English merchants established a number of small fishing camps along the New England coast and on the nearby islands. For the next two decades, fishermen from the western coast of England dominated the fishing business. They shipped most of their catch to England, but sold part of it to resident white settlers. With the outbreak of the English Civil War in 1641, the Atlantic fishing industry underwent a significant change. Warfare interrupted commerce, making it difficult for English merchants to dispatch fishing fleets. The temporary absence of English fishermen opened up the fishing grounds to New England settlers. They began to catch the fish they formerly had to buy. According to Governor John Winthrop, the 1641 catch amounted to some 300,000 cod.

The fishing industry soon became one of the mainstays of the New England economy. In 1721, the Reverend John Wise marveled that Gloucester, a fishing village first settled in 1623, originally "was but a Rock . . . Yet by Merchandise became the Queen of the Sea, the Metropolis of the World." Wise's obvious exaggeration aside, Gloucester and other coastal fishing villages did experience a significant economic and social transformation in the eighteenth century. They became stable and prosperous communities, better off than many of their agrarian counterparts in the interior. On the eve of the American Revolution, fish accounted for more than one-third of the volume of New England's exports, and as many as 10 percent of the region's adult males worked full or part time on the fishing boats.

The expanding fishing industry generated business for New England merchants. They purchased and distributed the catch, and sometimes even outfitted and sponsored fishing expeditions. More important, they also gained a share of the foreign fish market. New England merchants exchanged dried fish for manufactured goods offered by London merchants, who in turn traded the fish for wine produced in southern Europe and the Wine Islands (the Azores, the Canaries, and the Madeira). As Chart 3.3 on p. 70 shows, English imports into New England followed a general upward trend in the eighteenth century. Moreover, New England merchants developed additional sources for trade elsewhere. Some merchants began to trade directly with the Wine Islands, Portugal, and Spain, dealing not only in fish, but also in wooden staves for making wine casks. Merchants also traded extensively with the sugar islands of the West Indies, acquiring sugar and molasses for consumption in New England. These energetic traders extended their shipping and marketing network to both sides of the Atlantic. They went virtually everywhere and traded in virtually everything. This basic pattern of trade, established in the middle of the seventeenth century, remained essentially unchanged up to the time of the American Revolution.

CHART 3.3
Population and Imports in New England, 1700–1780

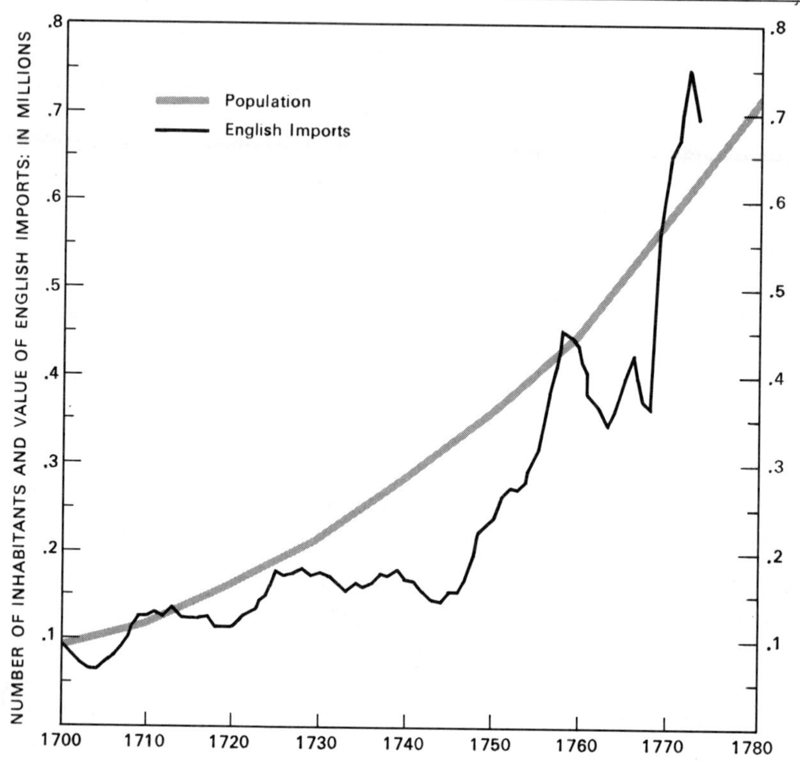

SOURCE: *Historical Statistics*, Series Z 3–8; John J. McCusker, "The Current Value of English Exports," *WMQ* 28 (1971), Table III.

In New England, the low level of English imports before 1750 was due primarily to a lack of valuable export commodities. The subsequent rise in imports reflected rising profits from exports of corn, meat, and wood products to the West Indies and from shipping and other commercial activities.

Economic Expansion and Urban Development

Trade transformed New England, especially the seaport towns. By the end of the seventeenth century, Boston's population had risen to 6000, making it by far the largest urban center in the American colonies. Its merchants controlled 40 percent of the carrying capacity of all colonial-owned shipping. These merchants naturally exerted considerable influence on the political and social life of the region. Their unquestionable contribution to New England's growing wealth gave the more prominent of them a stature equal to that of the leading Puritan ministers and magistrates.

Some people feared that this growing commercial wealth would undermine the purity of the Puritan commonwealth. Every merchant ship that docked in

New England's harbors brought new people, many of whom had no commitment to the original purpose of the Puritan mission. The same ships also brought new goods and customs, some of which shocked the sensibilities of pious Puritans. Reverend Increase Mather lamented:

> A proud Fashion no sooner comes into the Country, but the *haughty Daughters of Zion* in this place are taking it up, and thereby the whole land is at last infected. What shall we say when men are seen in the Streets with monstrous and horrid *Periwigs*, and Women with their *Borders* and *False Locks* and such like whorish fashions, whereby the anger of the Lord is kindled against this sinful Land!

The attractions of commerce proved extremely powerful to an increasing number of urban dwellers. At the end of the seventeenth century, one of every ten adult males in Boston described himself as a "merchant" on the shipping records; one of every six males had a share in a maritime venture. Most of these men were small traders, shopkeepers, millers, artisans, or farmers, who combined their scarce capital resources in order to enter the Atlantic trade.

Even those who did not invest directly in shipping still felt the effects of the expanding economy in the first half of the eighteenth century. The demand for vessels to transport goods sparked the appearance of dozens of small shipbuilding yards. These shipyards employed carpenters, rope and sail makers, caulkers, and ordinary laborers. The large amount of wood required by shipbuilders created an increasing number of jobs for lumbermen, sawmill operators, and wood merchants. Other merchants and artisans began new enterprises to process imported goods for local distribution or to produce goods for the export trade. As Thomas Bannister noted in 1714 in his *Essay on the Trade of New England*, the Atlantic trade

> employs a great number of Ships and sailors. Tradesmen feel the Benefit by the Merchandise of Soap, Candles, Beer, building of Ships and the great Number of Casks this Trade employs. The landed Interest Shares with them in the Export of very much Hay, Oats, Onions, Apples, Pork, Beef, Staves, board, butter, and Flour. The Fishery by a great Export of Mackerel and refuse cod. . . . The return for these is Molasses, which we brew and distill, and thereby raise many good Livings.

The appearance of colonial refineries reduced the need for imported English loaf sugar after 1750. And two decades later, 1.1 million gallons of rum per year were being distilled in Boston and exported to markets once served by the rum producers of the West Indies.

Members of the urban merchant class enjoyed the greatest gains. They made a handsome profit not only on foreign trade, but also on colonial wars. During the period of conflict with Spain and France (1739–1746), and during the latter part of the French and Indian War (1754–1763), England pumped large amounts of money into the colonies to pay for new ships and supplies for troops. Most of this money passed through the hands of urban merchants. Some merchants also outfitted privateering vessels and gained substantial profits from the legalized piracy of enemy ships. During both war and peace, Boston's wealthy elite became even wealthier. In 1687, the top 10 percent of the city's taxpayers owned 46 percent of all taxable property; in 1771, their share had increased to 63 percent.

The benefits of economic growth did not accrue to other urban dwellers. After 1740, the vast majority of Boston's inhabitants saw their economic vulnerability increase. The colonial wars that brought infusions of British funds also brought higher taxes and prices. Between 1734 and 1749, the costs of war caused the average Boston taxpayer's bill to increase fourfold, from 12 to 50 shillings. The price of bread rose at about the same rate. People on fixed incomes — clergy, schoolteachers, and many widows — suffered a disastrous decline in their buying power. Most urban laborers found that the rise in wages did not keep pace with the increasing cost of living. Moreover, few workers enjoyed full employment. One ship carpenter near Boston recorded that he worked only 296 days in a period of two years. Semiskilled or unskilled workers faced competition from both newcomers and slaves as well. In 1742, Boston's 1374 slaves represented 8.5 percent of the total population. Hundreds of the city's inhabitants had virtually no property and no prospects. According to the Boston selectmen, even the "Middling Sort of People are daily decreasing, many of them sinking into extreme Poverty." The large share of the wealth that was controlled by Boston's elite in 1771 reflected an extreme economic inequality that had been prevalent for decades.

The Middle Colonies: Grain Exports and Economic Growth

After 1750, the middle colonies emerged as the most economically dynamic and dominant region in British North America. The pattern of development generally followed that of New England but also partly reflected the southern economy. The economic base of the middle colonies combined the best features of the other regions — extensive commercial activity with staple crop production. The fertile farmlands of Pennsylvania, New Jersey, and New York's Hudson Valley produced an agricultural abundance unmatched anywhere else on the mainland. Wheat production in the middle colonies made them the breadbasket of North America, suppliers of grain to Europe and the West Indies. But unlike the southern colonies, the middle colonies did not restrict themselves to a monoculture economy, tied to the vagaries of a single market and almost wholly dependent on imported goods. They developed a diversified economic base in which local merchants supplied and directed the capital for a wide range of enterprises. By the 1770s, New York and Pennsylvania were the two most successful American colonies.

As in New England, the process of development was most evident in the cities. New York City and Philadelphia lagged behind Boston in both population and commerce in the seventeenth century. In 1690, New York's population of 4700 came close to Boston's 6000, but Philadelphia was still a relatively new town with only 2200 inhabitants. As well as being more populous than Philadelphia, New York had by far the more active commercial life. Its location at the mouth of the Hudson River gave it direct and easy access to the interior. For that reason, the original Dutch settlers had made it the center of their colonial trade. Even after the takeover by the English in 1664, New York City remained economically dominated by Dutch inhabitants; as late as 1695, its 34 English merchants were outnumbered by 46 Dutch traders. And yet compared to Boston, New York City and Philadelphia carried on only a small transatlantic trade. Even by the second decade of the eighteenth century, the total tonnage of ships clearing Boston was almost twice that from the other two ports combined.

The middle colonies experienced remarkable growth in the first half of the eighteenth century. A massive and diverse influx of European migrants — English, Scots, Irish, Scotch-Irish, and Germans — swelled the population of the region by more than 500 percent. The creation of thousands of new farms resulted in the production of a huge supply of grain, far more than the people of the middle colonies required for home consumption. At the same time, the rising demand for foodstuffs in Europe and the West Indies created an ever expanding market. Between 1720 and 1750, the price of wheat in the Philadelphia market increased by more than 50 percent; this upward trend continued until the value of wheat in 1770 was double what it had been fifty years earlier.

The success of the grain trade also helped stimulate other exports. One Philadelphian noted in 1741:

> We make our Remittances a great many different ways, sometimes to the West Indies in Bread, flour, Pork, Indian Corn, and hogshead stave, sometimes to Carolina and Newfoundland in Bread and Flour, sometimes to Portugal in Wheat, Flour and Pipe Staves, sometimes to Ireland in Flax Seed, Flour, Oak and Walnut Planks and Barrel Staves, and to England in Skins, Tobacco, Beeswax, Staves of all Kinds, Oak and Walnut Planks, Boat Boards, Pig Iron, Tar, Pitch, Turpentine, Ships, and Bills of Exchange.

The profits from this trade brought increasing prosperity. In some of the long-settled agricultural regions, such as Chester County in Pennsylvania, the earnings from wheat enabled farmers to subdivide their property among their children and still maintain the economic viability of the family farm. Moreover, part of the earnings from wheat could be used to pay for imported goods. The value of English imports sold in the middle colonies increased by 400 percent between 1745 and 1760, far outpacing the rate of population growth (see Chart 3.4 on p. 74). More important, the increase in trade led, as it did in New England, to the expansion of other sectors of the domestic economy. The high prices paid to thousands of independent farmers not only increased the demand for English goods, but also stimulated the growth of rural-based manufacturing. In

CHART 3.4
Population and Imports in the Middle Colonies, 1700–1780

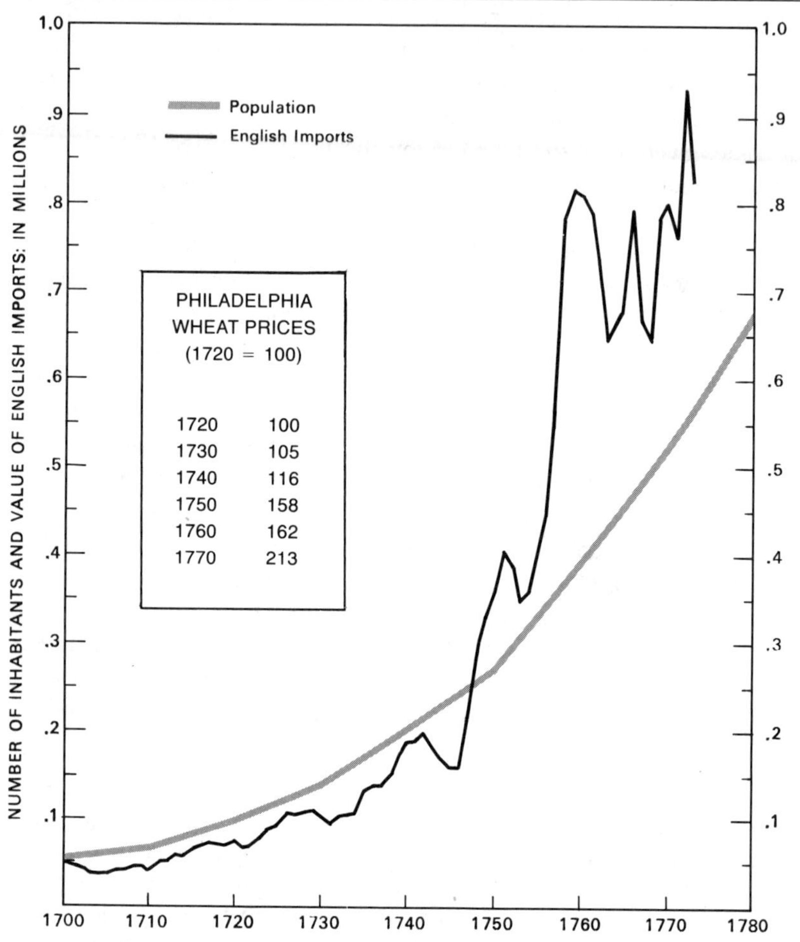

SOURCE: *Historical Statistics,* Series Z 9–11, 336, and 338; John J. McCusker, "The Current Value of English Exports," *WMQ* 28 (1971), Table III.

In the middle colonies, the sharp upsurge in imports after 1750 paralleled the growth of the wheat trade. Because of high demand, wheat prices doubled in England and America between 1720 and 1770. The increasing value of exported wheat financed the rising level of consumption in the middle colonies and led to the emergence of Philadelphia as the largest American port.

the cities, the shipbuilding and construction trades flourished; by the 1770s, Philadelphia's shipbuilding industry equaled that of Boston.

During the third quarter of the eighteenth century, the growth of the middle colonies was so spectacular that the region challenged New England for commercial supremacy. The leading New England colony, Massachusetts, could not even provide food for its own people; between 1768 and 1772, it had to

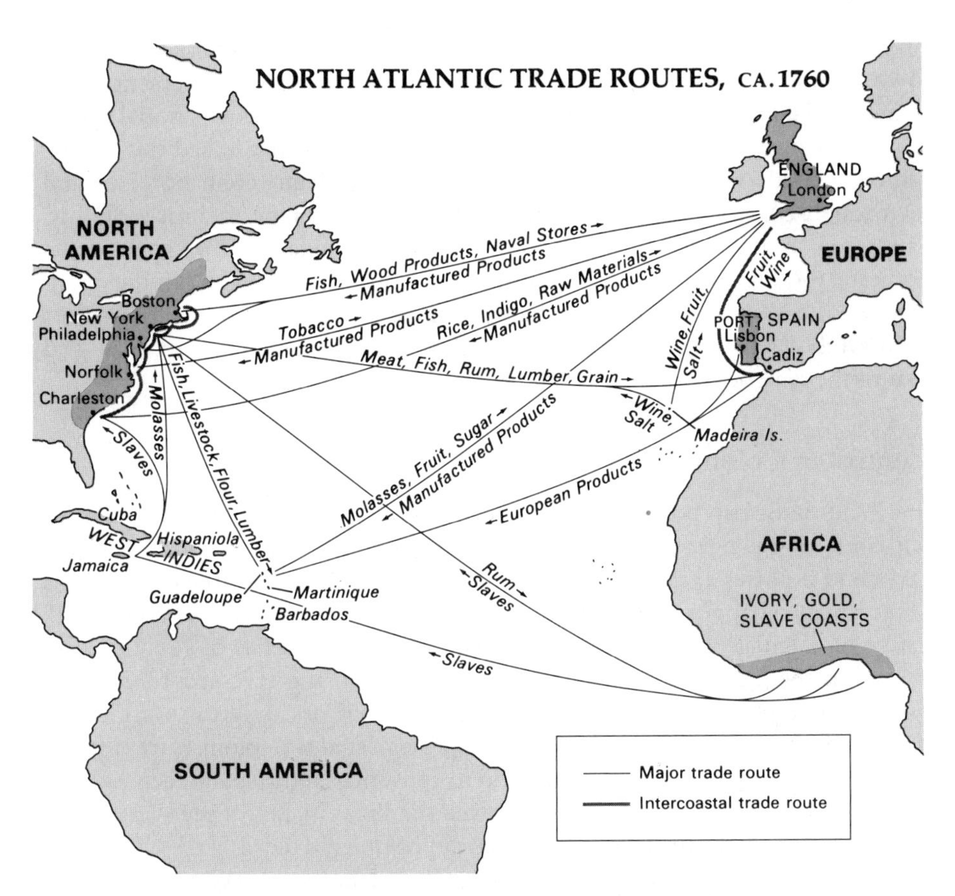

British and American merchant vessels criss-crossed the Atlantic Ocean, carrying agricultural and manufactured goods, European migrants, and African slaves. American vessels dominated in the trade between North America and the West Indies, and British firms controlled most of the other routes.

import about 14 percent of its basic foodstuffs from other colonies. In the same period, the food supply in New York and Pennsylvania exceeded the population's needs by about 30 percent and 50 percent respectively, and both colonies were able to export huge quantities of foodstuffs. Moreover, their shorter distance (and therefore travel time) to the West Indies gave New York and Philadelphia merchants an advantage over their Boston counterparts, and they began to cut into the New England merchants' domination of the intercolonial trade. The economic fortunes of the cities also contributed to different rates of population growth. On the eve of the American Revolution, Boston's population was still 16,000, about what it had been in 1740. In comparison, New York had grown to 22,000 inhabitants, and Philadelphia topped all American cities with a population of 25,000.

The significance of the three major northern cities far exceeded the size of their populations. Together they accounted for less than 3 percent of the North

American population in 1770. (In fact, the combined population of *all* colonial towns having more than 2500 inhabitants represented less than 7 percent of the total colonial population.) Still, the larger northern cities were the nodal points in an extensive and remarkably expansive trade network that linked the American countryside with European markets. Urban merchants contributed to and increasingly controlled complex commercial operations in the Atlantic economy. This trade, in turn, provided much of the capital that financed the expansion of domestic production and construction, in the cities themselves and in the countryside. The economic activity of the major cities of the northern colonies thus both symbolized and stimulated the development of the entire region — and that of the American colonies as a whole.

Control of Colonial Commerce

In 1775, the British politician Edmund Burke observed that "the settlement of our colonies was never pursued upon any regular plan; but they were formed, grew, and flourished, as accidents, the nature of the climate, or the dispositions of private men happened to operate." Burke clearly exaggerated the "accidental" nature of colonial growth and was mistaken about the absence of a regular policy. From the earliest days of colonial settlement in the seventeenth century, English government officials and merchants wanted not only to promote the colonial economy, but also to control it. They expected the colonists to produce valuable foodstuffs and raw materials and to consume English manufactured and finished goods. For that reason, they viewed the sugar islands of the West Indies as the prized possession among the American colonies. They looked with growing concern, moreover, on the rise of an aggressive merchant class and the emergence of local manufacturing, especially in the northern colonies. During the eighteenth century, the British government exercised varying degrees of control over the colonies, but its basic policy remained that of keeping the colonies economically subservient by restricting the development of independent economic activities.

The creation of the Board of Trade in 1696 underlined Parliament's continuing commitment to colonial regulation. Parliament acted because of military and economic concerns. From 1689 to 1713, England was almost continually at war with France, and the board played an important role in planning military operations against the French in Canada. The board also had extensive influence over colonial administration. It advised the Privy Council on the appointment of colonial governors and other officials, and it reviewed acts of the colonial assemblies to ensure that they did not conflict with the economic policies of the British government.

Following the creation of the Board of Trade, Parliament passed a series of acts that increased the number of American products subject to the board's regulation under the Navigation Acts. Rice and naval stores, for instance, were added to the list of enumerated commodities that had to be exported only to England. (The board exempted rice in 1729.) The Wool Act (1699) and the Hat Act (1732) prohibited the export of certain textiles and finished hats from the colonies altogether.

One of the most important — and revealing — pieces of English legislation was the Molasses Act (1733). The stated goal of this act was to protect West Indian sugar producers who "have of late years fallen under such discouragement, that they are unable to improve or carry on the sugar trade upon an equal footing with the foreign sugar colonies." Accordingly, Parliament imposed a duty on any sugar product imported from outside the English possessions. West Indian sugar planters, who had lobbied strenuously in Parliament for such protection, were delighted with the legislation. New England merchants, who often imported molasses from the French West Indies, were not. Parliament's obvious favoritism toward West Indian agricultural producers to the detriment of mainland merchants reflected the fundamental assumption of mercantilist policy: the colonies should contribute to, but not compete with, Britain's growing dominance of the Atlantic trade.

In practice, however, neither the Molasses Act nor any other parliamentary regulation imposed serious restrictions on colonial economic development. Although the Board of Trade attempted to bring colonial customs officers more directly under its authority, officials in London could do little to control the actual movement of goods in and out of American ports. New England merchants generally ignored the Molasses Act and resorted to smuggling and bribery to evade the duties. Distance and deceit made colonial trade regulations virtually unenforceable.

Changing military conditions also affected colonial economic policy. After the Treaty of Utrecht in 1713 ended the war with France, British policy makers had less reason to fear for the safety of the northern mainland colonies. Ensuing conflicts with Spain shifted the military focus to the Caribbean and, to a lesser degree, to the southernmost mainland colonies. For that reason, Robert Walpole, who as First Lord of the Treasury and Chancellor of the Exchequer was primarily responsible for British foreign policy throughout most of the period from 1714 to 1742, paid comparatively little attention to the mainland colonies. Walpole and his Secretary of State for the Southern Department, Thomas Pelham-Holles, the Duke of Newcastle, adopted a policy of "salutary neglect." They initiated virtually no new policies, content to let the colonial system operate without a great deal of official interference. Even if the colonial system worked imperfectly, it still worked for Britain's benefit. Many American colonists, especially northern merchants, felt the system worked for their benefit as well.

The period of neglect, however salutary, was short-lived. With the recurrence of war with the French in the 1740s and again in the 1750s, British officials redirected their attention to the mainland colonies. In 1748, the Earl of

Halifax became president of the Board of Trade, and he undertook a much more aggressive program of colonial regulation. From then until the outbreak of the American Revolution, successive administrations attempted to impose greater control over the colonial economy. Eventually the results would prove disastrous. As British officials found to their dismay in the 1760s and 1770s, the anticipated gains from taxes and customs revenues could never offset an even greater loss — that of the colonies themselves.

4

The Social Synthesis: Patriarchy, Property, and Power in Eighteenth-Century America

The most important authority governing people's lives does not always come from distant rulers. Control is also exerted by sources much closer to home, sometimes literally within the home itself. The day-to-day, face-to-face relationships that people quite often take for granted involve subtle assumptions (and assertions) of power. Many of the basic differences that distinguish people from one another — whites from blacks, men from women, old from young, rich from poor — also define inequalities of status and authority. To a large extent, these forms of inequality become institutionalized in law, and thus are supported by the government. Just as often, however, they pervade the culture in less formal ways; they are sustained by custom, coercion, and even consent.

So it was in the colonies. Although Great Britain had become a significant global power by 1700, it still did not — indeed, could not — exercise constant control over its American subjects. Three thousand miles of ocean made regular communication difficult and rigorous command almost impossible. Far more important than the British government in the lives of most colonists were their local institutions — the representative assembly, the county court, the church, the town meeting. Magistrates, ministers, and other local leaders stood close at hand, highly visible figures embodying power through both position and personal prowess. Yet even they exercised their authority only on official, almost ritual, occasions.

In the American colonies, the family was ultimately the center of authority. It provided the primary means of social control for adults as well as children. It also defined a person's identity within the community. The ability to draw on family resources, especially inherited property and kinship connections, greatly aided one's achievement of status and power. The nature and norms of family life differed throughout the colonies, depending on regional culture, social class, and ethnic background. Yet everywhere the family fulfilled the essential function of socializing individuals, of integrating them into the broader context of society.

Parental Control in the Puritan Family

No group of Americans believed more strongly than the New England Puritans that authority begins at home. Unlike the Chesapeake area, which had attracted a significant proportion of young, single people, New England had been settled by people who migrated for the most part in household groups. From the first days of settlement, the family became, both by prescription and in practice, the main building block of the new society. Ministers often referred to the family as a "little commonwealth," a model for the larger society. Puritans were convinced that only stable, well-governed families could provide the firm foundation for a stable, well-governed social order. As Cotton Mather, Boston's most prominent minister, put it, "Families are the nurseries for church and commonwealth — ruin families and ruin all."

The Puritan emphasis on family government reflected the importance of children in Puritan society. Young children created an obvious burden for parents, especially for mothers, but they provided important benefits as well. In economic terms, a large family could be critical to survival. In their younger years, children provided a valuable pool of free farm labor, and in their later years, an important source of support for aging parents. In psychological terms, children represented a source of continuity that transcended age itself. Naming patterns indicate that parents consciously chose to pass their own names (and the names of earlier family members) down to children. Family identity, like family property, represented an important form of inheritance. The seventeenth-century Puritan poet Anne Bradstreet perhaps best expressed a parent's desire for immortality in a poem she wrote to her eight young "birds":

> When each of you shall in your nest
> Among your young ones take your rest,
> In chirping language oft them tell
> You had a Dam that lov'd you well.
> That did what could be done for young,
> And nurst you up till you were strong,
> And fore she once would let you fly,
> She shew'd you joy and misery;
> Taught what was good, and what was ill,
> What would save life, and what would kill.
> Thus gone, amongst you I may live,
> And dead, yet speak, and counsel give.

Children represented the "rising generation" that would shape the future not only of the particular family but of the society as a whole. For that reason, Puritans raised their children in a psychological environment that emphasized the individual's subjection to God, community, and family. By the time children reached the age of 6 or 7, they were dressed like little adults and were trained to take on the occupational and social roles they would have as adults. Boys joined their fathers in the fields or in the artisan's shop. Girls remained in the house with their mothers to learn domestic skills. Equally important, parents admonished each child to obey the moral code of the culture. There could be no dichotomy between self and society. John Robinson, the minister of the original Plymouth settlement, argued in an essay titled "Children and Their Education":

> And surely there is in all children . . . a stubbornness, and stoutness of mind arising from natural pride, which must, in the first place, be broken and beaten down; that so the foundation of their education being laid in humility and tractableness, other virtues may, in their time, be built thereon.

The very process of education reflected Robinson's philosophy. Even before children could read, they memorized commandments and catechisms. Catechisms focused special attention on the fifth commandment — "Honor thy father and thy mother" — because that commandment spoke directly to the duties of children toward their elders. In the context of Puritanism, the fifth commandment implied a set of hierarchical relationships that extended far beyond the immediate family. One catechism pursued the issue:

> Q. Who are here meant by Father and Mother?
> A. All our Superiors, whether in Family, School, Church, and Commonwealth.

When children were taught to read, the goal was not simply to help them master a useful skill, but to make them better Puritans. A Massachusetts law of 1647 required towns to establish reading schools because, as the General Court warned, "one chief project of that old deluder, Satan, [is] to keep men from the knowledge of the Scriptures." This Puritan commitment to Bible reading resulted in a remarkably high literacy rate. The percentage of men able to read and write rose steadily through the seventeenth and eighteenth centuries, so that by 1790 around 85 percent of all adult males in New England were literate, compared to around 60 percent in England. (The literacy rate for women, who received less formal education, was about half that for men.)

The connection between literacy and religious orthodoxy began even before a child could undertake serious Bible study. The basic text for beginning readers, the *New England Primer,* in its first lesson on the first letter taught about the first sin: "A: In Adam's fall, we sinned all." Young Puritans went on to read a short lesson in religion, always rendered in rhyme for easy memorization, for each letter of the alphabet. The goal of religious education was first to turn the young inward, making them anguish over their sinfulness and depravity

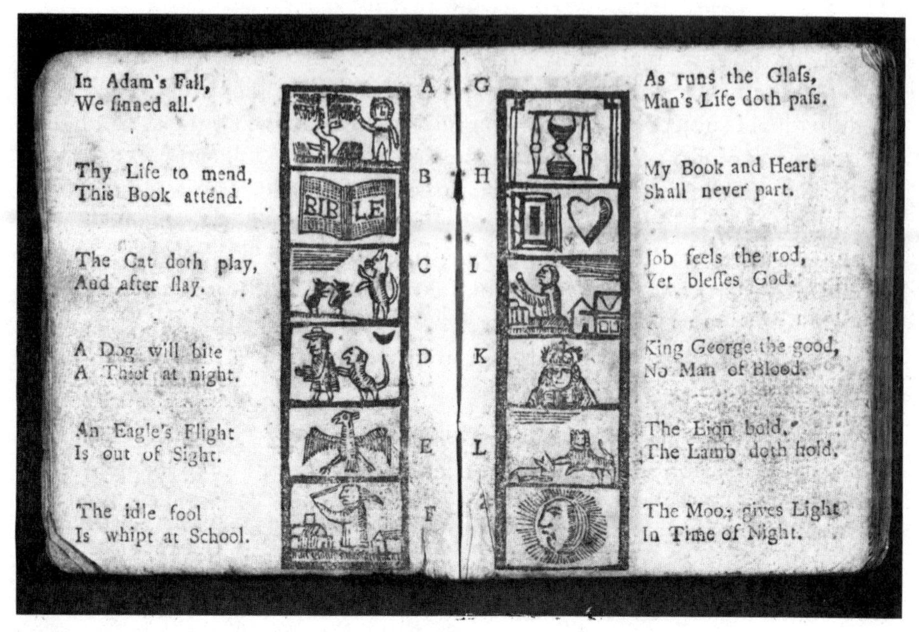

A
In Adam's Fall,
We ſinned all.

B
Thy Life to mend,
This Book attend.

BIBLE

C
The Cat doth play,
And after ſlay.

D
A Dog will bite
A Thief at night.

E
An Eagle's Flight
Is out of Sight.

F
The idle fool
Is whipt at School.

G
As runs the Glaſs,
Man's Life doth paſs.

H
My Book and Heart
Shall never part.

I
Job feels the rod,
Yet bleſſes God.

K
King George the good,
No Man of Blood.

L
The Lion bold,
The Lamb doth hold.

The Moon gives Light
In Time of Night.

Pages from The New England Primer. The short, rhyming verses of this early spelling book, first published in the 1680s, taught children the basic moral precepts of Puritanism along with the alphabet. (*Source: American Antiquarian Society.*)

as human beings, and then to turn them outward, making them recognize their need for external guidance.

Parents played the most crucial role in establishing standards of social control. As one minister explained, "When Minister's work in public is done, the Family work begins." Puritan ministers prescribed no single method of parental discipline. Some parents resorted to beatings and other forms of corporal punishment. Cotton Mather warned, "Better whipt, than Damned." But Mather himself sought to discipline his own children without physical abuse. In his *Bonifacius: Or Essays to Do Good* (1710), Mather expressed his determination "to avoid that fierce, harsh, crabbed usage of my children, that would make them dislike and tremble to come in my presence." "I would treat them," he continued, "so that they shall fear to offend me, and yet heartily love to see me." Rather than use physical coercion and humiliation, Mather consciously withheld or bestowed affection as a method of discipline. He explained his method in another essay:

> I first beget in them a high opinion of their father's love to them, and of his being best able to judge, what shall be good for them. Then I make them sensible, tis a folly for them to pretend unto any wit and will of their own . . . my word must be their law. . . . I would never come to give a child a blow; except in the case of obstinacy or some gross enormity. To be chased for a while out of my presence I would make to be looked upon, as the sorest punishment in the family.

This sort of parental discipline heightened the moral anxiety already inherent in children's religious education. Children had to accept their parents' (and their culture's) definition of what was right and wrong. They also had to accept responsibility for their own actions. Children thus learned to avoid parental (and social) disapproval through continual self-reform as well as external direction.

When Puritan children became adults, this inner psychological dynamic had important social implications. It bred a compulsiveness that could be harnessed only by a strict organization of activity. "Tis not *Honest,* nor *Christian,*" wrote Cotton Mather in 1701, "that a *Christian* should have no *Business* to do." Those who succeeded in their business — whether on the family farm or in some entrepreneurial activity — could look on their growing wealth as the reward for virtuous industry. They could not look on that wealth as something to enjoy, however; they had the obligation to reinvest it for the benefit of their children and their society. Moreover, those who did not succeed economically had to face the shame and frustration of failure — not just personal, but familial and social as well.

Parents and Property

Parents provided for their children materially as well as spiritually. Puritans practiced partible inheritance, the division of family property among all surviving heirs. Following a Biblical injunction, they accorded an extra share to the eldest son. Parents' responsibility to provide productive property for all their children placed a significant economic and emotional burden on them.

The inheritance system greatly reinforced parental authority. A father often used the promise of property to delay, or to determine, the timing of a son's marriage; in the meantime the father kept the young man at home as a source of free labor on the family farm. When a son did marry (usually in his mid-twenties), he did not always become an autonomous householder. Sometimes a father continued to hold title to the land, allowing his son to farm it, build on it, even raise a family on it, but not to own it. The father's death did not always give a son full control of the land. A will often specified how the heir must use the property or live his life. When John Lovejoy of Andover, Massachusetts, wrote his will in 1690, he left the family homestead to his youngest son, Ebenezer, on the condition that the boy look after his mother as long as she remained a widow. She was to live in the family's house and have a supply of food, a milk cow, a horse to ride to church, money for personal expenses, and "Suitable help in Sickness or weakness." Death brought about the division of the paternal property, but it did not always bring an end to paternal authority.

The Puritan emphasis on family order placed a strong cultural compulsion on young adults to marry. Most parents took the responsibility for choosing their children's mates. Young people might veto prospective mates they found personally unacceptable, but they could not presume to make a marriage decision on their own. Throughout the eighteenth century, most young women in

New England families married in the order of their birth; parents generally did not allow a younger daughter to marry until they had made suitable arrangements for any older sisters. Marriage was not simply an emotional union based on mutual love between a woman and a man; it was also an economic arrangement that affected both their families. Parents provided land for sons and movable property (a dowry) for daughters, and thus had a tangible investment in their children's adult lives.

Woman's Place in Puritan Society

The transition from daughter to wife by no means freed a woman from her dependent status. A woman lived virtually her whole life subject to the authority of a man, either her father or her husband. According to common-law doctrines transferred from England, a married woman did not exist as an autonomous individual; in most cases, she could not own property, make contracts, or file suit. As the English jurist William Blackstone explained in 1765, "In marriage, the husband and wife are one person in law; that is, the very being or legal existence of the woman is suspended during the marriage, or at least is incorporated or consolidated into that of the husband."

The standards of Puritan society made it quite clear that the husband and father was to rule within the "little commonwealth" of the family. His wife, no less than his children, owed him deference and obedience. In *The Well-Ordered Family* (1712), the Puritan minister Benjamin Wadsworth spelled out the mutual obligations between husband and wife:

> Though the Husband is to rule his Family and his Wife . . . he must not treat her as a Servant, but as his own *flesh:* he must love her as himself. He should make his government of her, as easy and gentle as possible; and strive more to be lov'd than fear'd; though neither is to be excluded. On the other hand, Wives ought readily and cheerfully to obey their Husbands. *Wives submit your selves to your own Husbands, be in subjection to them.*

Despite her inferior social and legal position, a wife played an important role in the raising of crops and children. Virtually everything the members of her family wore or ate was produced by her hands. In that respect, women made a significant contribution to the economic and emotional stability of the family. The Puritan minister Samuel Willard recognized the role wives shared with their husbands in governing the household:

> If God in his Providence hath bestowed on them Children or Servants, they have each of them a share in the government of them; tho' there is an inequality in the degree of this Authority, and the Husband is to be Acknowledged to hold a *Superiority,* which the wife is practically to allow; yet in respect to all others in the Oeconomical Society, she is invested with an Authority over them by God . . . for tho' the Husband be the Head of the Wife, yet she is an Head of the Family.

The courts of New England did not always apply the standards of the common law with total strictness or inflexibility. In seventeenth-century Plymouth Colony, courts occasionally upheld the right of women to make contracts after marriage. A wife's right to an inheritance was also protected. Strict legal rules guaranteed a decent provision for widowhood, regardless of any irresponsible or irrational actions of a husband or his heirs. Court records also reveal numerous cases in which women resisted the authority of their husbands. At the Salem Quarterly Court in 1678, Bridget Oliver was "presented for calling her husband many opprobrious names; as old rogue and old devil." She was ordered "to stand with her husband, back to back, on a lecture day in the public market place, both gagged for about an hour with a paper fastened to each of their foreheads, upon which their offence should be plainly written." Sometimes public humiliation alone could not reconcile private differences. In instances of abusive treatment or desertion, women often came before the court to seek protection. Divorce, however, was rare. Massachusetts courts received only 229 divorce or separation petitions between 1692 and 1786, and 69 percent of those came after 1755. More than half of those petitions (128) were filed by wives.

Clearly the "little commonwealth" had its share of conflict. The husband and father who occupied the top of the household hierarchy was not immune from challenge. Still, the Puritan prescriptions for family government created strong legal and cultural imperatives that made change difficult.

Parents and Children in Quaker Pennsylvania

Quaker families resembled Puritan families in their basic structural features. Quaker men in their mid-twenties normally married women in their early twenties, and these unions usually produced seven children who lived to be adults. Husbands had authority over their wives, and parents exercised control over their children. But Quaker family relationships existed within a cultural context that differed from that of Puritan New England. Within the family itself, Quakers emphasized equality over hierarchy and gentle guidance over strict discipline. The church and community were the stronger sources of control.

Unlike the Puritans, who relied on the written word as found in the Bible, Quakers found "the Word" in "holy conversation." They gave spontaneous expression to their spiritual experience through human speech and interaction. The main forum for such expression was the weekly meeting, where all members of the community had equal opportunity to speak. Quakers took great care to make sure that all who participated in the meeting were spiritually acceptable, not tainted by excessive worldly concerns. For that reason, they exercised strong communal control over marriage. The 1694 Yearly Meeting of Philadelphia's Society of Friends warned parents to "take heed of giving your Sons and Daughters (who are Believers, and Profess, and Confess the Truth) in Marriage with Unbelievers for that was forbidden in all Ages, and was one Main Cause that brought the Wrath of God upon Old Israel." Those Quakers who married outside the sect faced expulsion. The goal was to keep everyone, parents and children alike, isolated from the carnal world and encompassed within the Quaker community.

To enable them to be stable members of the community, Quaker couples received generous gifts of land and property from their parents. For Quakers, a central purpose of accumulating property was to provide for one's children. To promote "the more convenient bringing up of youth," William Penn urged the creation of large farms in Pennsylvania. In the Chester and Welsh Tracts, two areas of early Quaker settlement, the average family holding in the 1690s comprised more than 330 acres. Moreover, many families increased their land holdings as the number of children grew. Sons of the first settlers received an average of over 200 acres from their fathers, and daughters received the equivalent in cash or goods. Indeed, most sons received more land than they needed. The large inheritance was the father's way of making sure that his descendants remained safe from the evils of the carnal world.

In spite of this child-centered system of property accumulation, Quaker fathers exercised less control than did their New England counterparts. Having instructed their children through personal example and exhortation, they did not attempt to govern their adult offspring through the inheritance system. They generally gave their sons outright gifts of land at the time of marriage, without attaching any restrictions or stipulations.

In comparison to Puritan children, Quakers had the advantage of growing up in a spiritually egalitarian and materially abundant environment. They too learned habits of self-discipline and frugality — traits that were important to economic success. These emotional and economic resources enabled Quakers to become remarkably prosperous in the broader context of Pennsylvania society. And even though their culture promoted personal success, it also demanded commitment to the community. For Quakers as well as Puritans, individual and familial autonomy were clearly submerged within a communal religious and social system.

The Social Implications of Inheritance

By the middle of the eighteenth century, the northern American colonies had undergone a complicated and even contradictory process of social evolution. In thousands of small towns and villages, bonds of kinship helped maintain a cohesive social unit and a strong sense of community. At the same time, the individual household had emerged as the primary economic and social unit. Important decisions were always made with reference to the good of the nuclear family. The consequences of those decisions, however, affected not only individual family members, but their communities and the society as a whole.

The practice of partible inheritance originally helped to preserve a relative degree of stability and equality, both in the family and in the rural community.

Guaranteeing all children a share of the family estate impeded the concentration of wealth in the hands of a few favored sons. In most farming communities, inheritance did not bring instant wealth. Young men usually accumulated wealth gradually as their years and their families increased. In eighteenth-century Guilford, Connecticut, all older heads of households were in the top 40 percent of wealth holders; the top 20 percent contained only a handful of young men.

The limits to landed inheritance increased over time. In seventeenth-century Dedham, Massachusetts, the average estate among first-generation settlers consisted of about 150 acres — more land than a family needed to supply its own needs. By the early eighteenth century, however, the typical third-generation inhabitant owned 100 acres, and over the course of that century average farm size in Dedham shrank to around 50 acres. The decline in the size of family land holdings made it increasingly difficult for fathers to give each son a farm large enough to support a family. A similar situation prevailed in nearby Andover, Massachusetts. Fully 95 percent of the estates of the first-generation settlers were divided among *all* the male heirs. Only 75 percent of the estates were so distributed by the second generation, and 58 percent by the third generation. The promise of inheritance varied from family to family. Some fathers prospered because of good luck or shrewd agricultural practice, and they acquired extra land to give to their children. Still, the overall trend was that an increasing number of sons could not expect to receive a landed estate.

Sons who did not receive land in their father's community often had to change vocation or location. Some became apprentices, at family expense, in the households of local artisans. A few sons from prosperous families attended college to prepare for careers as ministers or lawyers. Many others received a gift of money or a deed to land in a newly opened settlement. One Connecticut father wrote to a Massachusetts kinsman:

> As God has blessed me with a numerous off-Spring, and it Suiting my Affairs much better to bring my Boys up to husbandry than to put 'em out to Trades; but not having Land Sufficient for Farms for 'em all; I purpose . . . [to] purchase Some of 'em Settlements in Some new Towns where Land is good & Cheap.

Although many of these new towns lacked the religious or communal intensity of the early Puritan or Quaker settlements, the young settlers still established roots as farmers. Thus the traditions and customs of small-scale farming were propagated across the landscape as the population spread. Continuity in family choices contributed to the cultural continuity of northern rural society.

Death and Disorder in the Chesapeake Family

Throughout most of the seventeenth century, demographic conditions severely limited the development of stable families in the Chesapeake region. The high ratio of men to women made marriage more the exception than the rule for men; a majority of men simply could not find a wife. Moreover, the high mortality rate made marriages short; in one Maryland county, half the marriages ended within seven years because of the death of one partner. Children likewise faced insecure lives. Almost one-fourth of the children in seventeenth-century Middlesex County, Virginia, lost a parent before the age of 5. By the time they reached 18 more than 70 percent had suffered the death of at least one parent, and around 30 percent were orphans. The responsibility for raising children often fell to adults — stepparents, godparents, or guardians — who had no blood connection to them at all.

The high rate of death in the early Chesapeake settlements greatly complicated family identity. Widows, widowers, stepchildren, and orphans abounded. Remarriages brought them all together in complex household arrangements that mingled the people and property of several families. Within these mixed households, lines of inheritance often became tangled, and often rather tenuous. For that reason, Chesapeake parents tried to make sure that their children's property did not fall into the hands of a greedy stepparent or guardian. In writing their wills, some fathers gave control of family land to sons still in their teens. Other men bestowed generous property settlements on their wives, specifying that the land be given directly to their children when they reached maturity. When widows remarried, they protected their children's inheritance by entering into detailed marriage contracts with their prospective husbands. In a society where few parents lived to see their children become adults, planning for the future was a complex and confusing process.

By the end of the seventeenth century, however, lower death rates brought greater stability to the family in the Chesapeake region. Parents lived longer. Property owners, especially of larger plantations, expanded their holdings and passed them to their children. In 1705, the top 10 percent of Virginia planters owned 55 percent of the patented land. During the eighteenth century, a handful of prominent and prosperous families established themselves as a strong and stable planter elite. Achieving continuity from one generation to the next in these gentry families represented a critical step in the social development of the South.

The Power of the Plantation Patriarch

Gentlemen planters took pride in their prominence. They fashioned a self-conscious social identity that defined the dominant cultural standards for all of southern society. Virginia's William Byrd II, one of the wealthier and more articulate members of his class, perfectly exemplifies this role. Born in America and educated in England, Byrd assumed the position of plantation master when his father died in 1705. He inherited the 6000-acre Westover plantation on the James River, plus another 20,000-acre holding farther upriver. During his lifetime, he expanded his property holdings to almost 180,000 acres of land and more than 200 slaves. In a letter to an English friend in 1726, Byrd wrote about his life with great contentment:

> I have a large family of my own, and my Doors are open to Every Body, yet I have no bills to pay, and half-a-Crown will rest undisturbed in my Pocket for many Moons together. Like one of the Patriarchs, I have my Flocks and my Herds, my Bond-men and Bond-women, and every Sort of trade amongst my own Servants, so that I live in a kind of Independence on every one but Providence.

Byrd no doubt exaggerated his economic independence. Like all wealthy planters, he relied on annual infusions of credit from English tobacco merchants to finance his plantation operations. To direct those operations, Byrd relied on no one but himself. He prized his independence and power.

Larger plantations like Byrd's produced more than tobacco. They also had grist mills, saw mills, textile works, iron works, tanneries, distilleries — all the facilities needed to make a plantation a self-sustaining community. A German visitor noted that such a plantation had "more the appearance of a small village by reason of [its] many small buildings." Planters had the assistance of their wives in governing the immediate household. Moreover, those planters who had outlying plantations used overseers to supervise production there. Running his plantation, Byrd wrote, "is attended with a great deal of trouble. I must take care to keep all my people to their Duty, to set all the Springs in motion and to make every one of them draw his equal Share to carry the Machine forward."

Westover Plantation in Virginia. This magnificent plantation house, constructed in the late 1720s by William Byrd II, typifies the Virginia gentry's self-conscious display of their wealth. The mansion was built in the Georgian architectural style, revealing an increasing emulation by prosperous Americans of English cultural models. (*Source: Library of Congress.*)

Personal authority was the key to an efficient plantation. The planter had to command respect and obedience. William Byrd's maintenance of his conjugal authority often involved the use of sheer physical force. He repeatedly ended arguments with his first wife, Lucy Parke Byrd, by overwhelming her sexually, or "flourishing" her. He noted in his secret journal in July, 1710: "In the afternoon my wife and I had a little quarrel which I reconciled with a flourish. . . . It is to be observed that the flourish was performed on the billiard table." Several months later, after arguing about Lucy's desire to pluck her eyebrows in the latest fashion, Byrd wrote that he "got the better of her, and maintained my authority."

Byrd's relations with his wife disclose the tension between affection and aggression that underlay family life among the southern gentry. Self-assured and assertive, planters placed great value on their personal independence and authority. They, and not civil officials, had the primary responsibility for preserving order on their plantations. Planters instilled an acceptance of duty and discipline among their dependents. Individual family members had to moderate their emotions for the sake of household harmony. If wives or children dared disrupt that harmony, masters would not hesitate to correct them with physical force. The masters' credibility as patriarchs on the plantations emanated outward from their immediate families.

Byrd and other slaveowners included their slaves within the patriarchal definition of their "people." Accordingly, they often adopted the language of family discipline — "correction" rather than "punishment," for instance — to refer to their treatment of their slaves. The masters' mental association of slave and family sometimes tempered the physical severity of discipline. But the planter took it as his right and duty to maintain strict control over everyone in his domain, both white and black. All were considered his subordinates, and none could escape his authority.

Relationships and Resistance in the Slave Community

Most slaves did not willingly accept their role as a master's property. They held on to their personal identities. Slaves came from a variety of African tribal backgrounds, as the map shows. Thrown together in the New World, they had to overcome their cultural diversity. Throughout the first half of the eighteenth century, thousands of newly imported blacks increased the slave population. Almost 90 percent of these new slaves arrived directly from Africa. The mixing of African-born and American-born blacks eventually created a distinctive Afro-American culture in the slave community. This common culture provided slaves with an important alternative — and even a form of resistance — to the power of their masters.

African-born slaves were the most resistant to assimilation into slave society. These "outlandish" slaves had known freedom before their captivity. After being transported to the New World, many sought to regain their liberty. The diversity of tongues and tribal cultures among them made organized collective

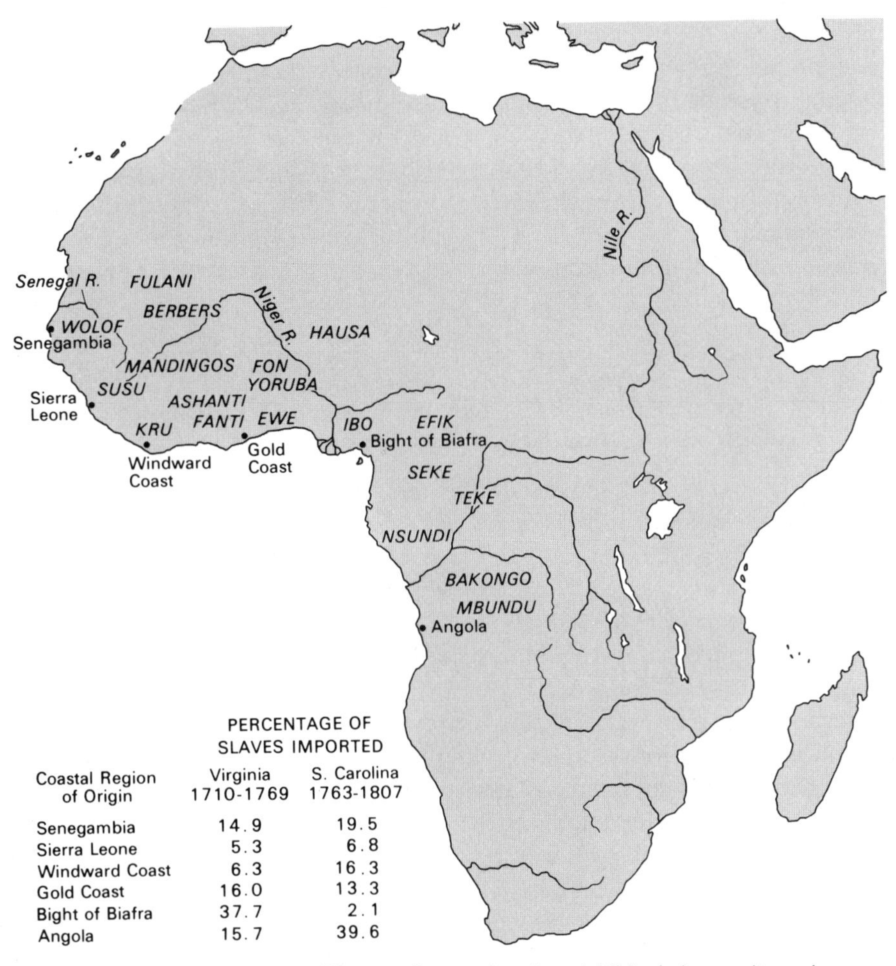

PERCENTAGE OF SLAVES IMPORTED

Coastal Region of Origin	Virginia 1710-1769	S. Carolina 1763-1807
Senegambia	14.9	19.5
Sierra Leone	5.3	6.8
Windward Coast	6.3	16.3
Gold Coast	16.0	13.3
Bight of Biafra	37.7	2.1
Angola	15.7	39.6

The importation of Africans from different tribes and regions inhibited slave unity and forestalled uprisings. The continuing influx of slaves during the eighteenth century reinforced the influence of African culture and delayed the emergence of a truly Afro-American society.

resistance extremely difficult. Occasionally members of a single tribe or language group stole boats and attempted to sail back to Africa. Others fled into the wilderness, where they tried to recreate the village society they had known in Africa. Their limited knowledge of the strange environment gave them little chance for success; most were recaptured — often by hired Indians — and returned to their masters. Some Africans were more fortunate. They found shelter with friendly Indian tribes and escaped slavery forever. These "maroon" communities of Indians and Africans represented a disturbing, somewhat mysterious, presence just beyond the reach of white society.

Some slaves turned to violent rebellion, despite its suicidal results. Blacks who struck or killed whites usually faced swift execution, but a number of small, organized uprisings still took place in the American colonies. The Stono Rebellion in South Carolina was the most important of these uprisings. On a Sunday morning in September, 1739, a group of slaves took up arms, apparently hoping to escape southward to Spanish-held Florida. They cut a path of death across St. Paul's parish, killing any whites they encountered. By the afternoon, the band of slaves (which never numbered more than 100) was surrounded by white militiamen and overwhelmed. Most of the rebels who did not die fighting were executed on the spot. Even though the Stono Rebellion was short-lived and unsuccessful, it fulfilled the worst fears of whites. A year after the uprising, one observer wrote:

> Such dreadful Work, it is to be feared, we may hear more of in Time . . . above all, if it is considered how vastly disproportionate the Number of white Men is to theirs: So that at best, the Inhabitants cannot live without perpetually guarding their own Safety, now become so precarious.

To avert a massive slave insurrection by the colony's black majority, South Carolina whites imposed strict controls over the actions and associations of their slaves. Throughout the South, the increasing stringency of slave laws clearly revealed the growing anxiety of slaveowners.

The vast majority of slaves, however, did not rebel or run away. Most slaves gradually formed familial relationships that tied them to the plantation. The initial experience of enslavement destroyed the traditional connections of African village life. Captivity tore parents from children, husbands from wives, kinfolk from community. Southern planters imported two or three male slaves for every female. This unevenness of gender distribution inhibited the development of slave families. By the middle of the eighteenth century, however, two factors created a more favorable context for stronger and more stable family ties. First, planters encouraged procreation to increase the slave population, and the native-born generation was, of course, equally divided as to gender. Second, the slave population became increasingly concentrated in larger plantation units. By the 1780s, almost 70 percent of the blacks in some tidewater counties of Virginia lived on farms of more than 10 slaves, as Chart 4.1 shows. Some plantations had well over 100. Those blacks living in a larger slave community experienced a richer personal and cultural life. Children had close kin — parents, grandparents, uncles, and aunts — to provide emotional support and role models. Often an *obeah,* a priest-like figure with a knowledge of witchcraft and other occult secrets, emerged as a spiritual (and political) leader in the slave quarters.

Masters often used the family ties among slaves for their own advantage. They recognized that family groups would be less rebellious than unrelated groups of young males. Adult slaves were much less likely to run away if they had to take their children along — or even worse, leave them behind. For that reason, many masters refrained from breaking up slave families through sale. On

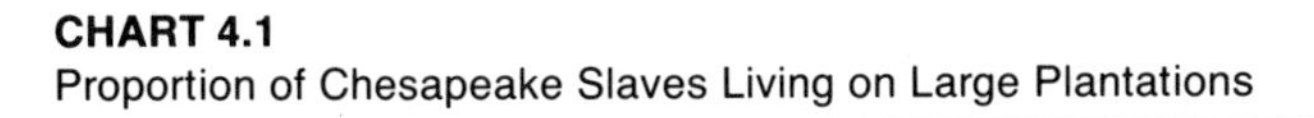

CHART 4.1
Proportion of Chesapeake Slaves Living on Large Plantations

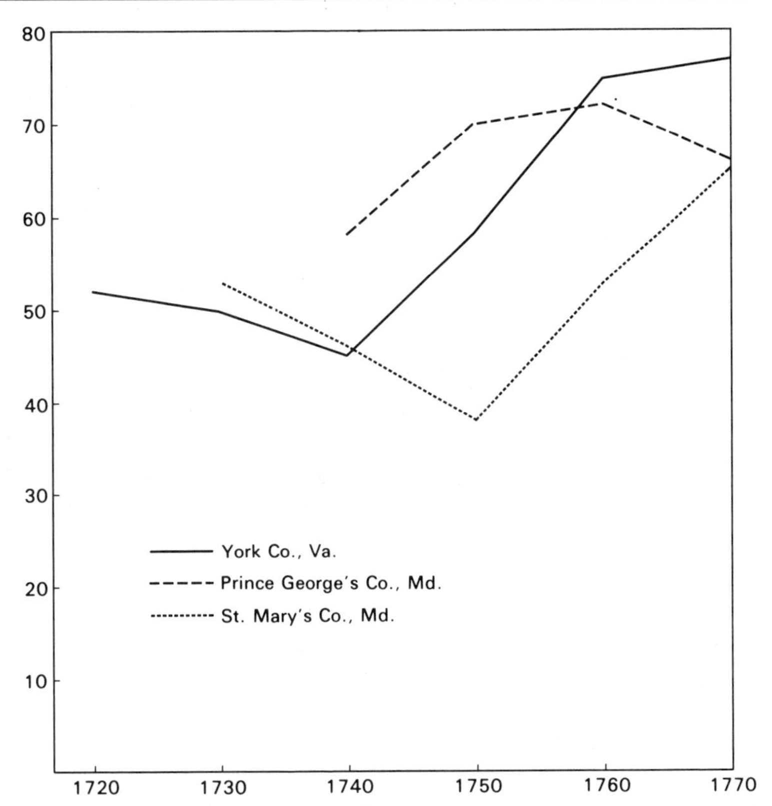

SOURCE: Allan Kulikoff, "The Origins of Afro-American Society in Tidewater Maryland and Virginia, 1700–1790," *WMQ* 35 (1978), Tables III and IV.

The growing number of slaves who lived on large plantations (those having more than ten slaves) reflects the concentration of land and labor in the hands of the wealthier planters in the Chesapeake region. During the eighteenth century, an increasing proportion of Chesapeake slaves were American-born, a sign of the greater opportunities for blacks to achieve settled family life and community ties on large plantations.

the other hand, some masters exploited black family ties by threatening to sell or punish a loved one.

Plantation slaves nonetheless found outlets for their rage. Some engaged in acts of subtle sabotage. They stole the master's food and liquor, broke his tools, and killed his animals. As one slaveowner complained, "[I]t is almost impossible to raise a stock of Sheep or Hogs, the Negroes are constantly killing them to sell to some white people who are little better than themselves." Slaves often shirked their responsibilities or intentionally worked at a slow pace. Many masters constantly complained that their slaves were indolent and incompetent, and some owners attributed the poor work habits of slaves to racial inferiority. Wiser

"The Old Plantation" (ca. 1744–1794). This anonymous painting suggests both the physical and the cultural separation between the slave quarters and the master's house. The headscarves, the musical instruments, and the dancing all reflect the African heritage that underlay the emerging Afro-American culture. (*Source: Abby Aldrich Rockefeller Folk Art Center, Williamsburg, Virginia.*)

owners saw in their slaves' behavior a pattern of subversion they were virtually powerless to prevent.

In the rice-producing lowlands of Georgia and the Carolinas, slaves asserted a measure of control over their work. Under the "task system," they labored industriously for the master, but they also claimed time to work for themselves. One observer explained this system:

> Their work is performed by a daily task, allotted by their master or overseer, which they have generally done by one or two o'clock in the afternoon, and have the rest of the day for themselves, which they spend in working in their own private fields, consisting of 5 or 6 acres of ground, allowed them by their masters, for planting of rice, corn, potatoes, tobacco, &c. for their own use and profit, of which the industrious among them make a great deal.

Slaves had a clear notion of what constituted a fair day's work for the master, and they considered their own free time as sacrosanct. Masters who attempted to impose a stricter work routine on their slaves did so at the risk of undermining the peacefulness and productivity of their plantations. "Should any owner increase the work beyond what is customary," warned one South Carolina planter, "he subjects himself to the reproach of his neighbors, and to such discontent amongst his slaves as to make them of but little use to him." Once the task system was established as a regional custom, slaves refused to surrender its prerogatives without resistance.

The master-slave relationship was never a simple or static one. It involved an almost constant struggle — sometimes physical, more often psychological — between the master's assertion of authority and the slave's aspirations for autonomy. The master had an enormous advantage: he could rely on the laws, resources, and even the armed might of white society to bolster his personal power. Slaves took advantage of what little leverage they had to gain concessions from their masters, and thus to gain some control over their own lives. Throughout the period of American slavery, neither the slaves' submission nor the masters' control was ever complete.

Planters in Politics

The power of the planter elite extended far beyond the plantation. Wealthy slaveowners used their personal prominence, family ties, and economic resources to assert control over the whole of southern society. Among whites, the social distance between gentleman and yeoman was immense — almost as great as that between master and slave. Members of the gentry emphasized this distance by making a pointed display of their wealth and power. Their dress and demeanor became potent symbols of social dominance. The son of a Virginia carpenter reported:

> We were accustomed to look upon, what were called *gentle folks,* as being of a superior order. . . . For my part, I was quite shy of *them,* and kept off at a humble distance. A *periwig,* in those days, was a distinguishing badge of *gentle folk* — and when I saw a man riding the road, near our house, with a wig on, it would so alarm my fears, and give me such a disagreeable feeling, that, I dare say, I would run off, as for my life.

Ceremonial occasions became ritual expressions of social identity and authority. Gentlemen justices dominated the sitting of the county court. Wearing wigs and fine clothing, they administered oaths and dispensed justice to their less wealthy neighbors. The imposing visual imagery complemented their legal authority. At church services, members of the gentry gave a dramatic display of their separate and superior status. As one observer noted, "It is not the Custom for Gentlemen to go into Church til Service is beginning, when they enter in a Body." Favorite gentry pastimes — horse racing, gambling, fighting, and dancing — became demonstrations of personal prowess. The gentry used every public event as an opportunity to project their power before the populace.

More important, the planter elite expanded its power, even as Chesapeake society grew in numbers and size. In Virginia, the total number of white households increased from about 6000 in 1700 to 80,000 in 1790, yet economic wealth and political power remained concentrated in the hands of the same family clans. Land and political leadership passed as a direct legacy from one generation to the next. As in the northern colonies, fathers practiced partible inheritance, but the division of family property did not mean privation for children of the gentry. Wealthier planters accumulated huge land holdings, usually spread out over two or more counties. As sons and daughters took up residence on their inherited lands, they extended their family's influence into new localities. Seven members of the Carter family, the richest of all the clans, owned a total of 170,000 acres of land and 2300 slaves scattered over seven different counties. Among the hundred wealthiest men in Virginia in 1789 were also nine members of the Cocke family, eight Fitzhughs, seven Harrisons, eight Lees, seven Randolphs, and eight Washingtons.

This small homogeneous elite — English by descent, Anglican in religion, and linked to one another by ties of kinship and economic interest — monopolized the political life of Virginia during the eighteenth century. Seven members of the Lee family represented five different counties in the Virginia assembly in the 1750s. In the decades preceding the American Revolution, no less than 70 percent of the 110 leaders of the House of Burgesses were drawn from great planter families resident in Virginia before 1690. That 60 percent of the adult white males in Virginia had sufficient property to vote was almost politically irrelevant. Given this concentration of hereditary power, the mass of poorer white planters deferred to the decisions of their "betters" and submitted grudgingly to their rule.

Family Prominence in Provincial Politics

In all the colonies, political power became increasingly concentrated in the hands of the elite during the eighteenth century. Intermarriage and economic alliance among prominent families created strong bonds of mutual interest and self-identity. From the local to the provincial level, the political culture was based on deference and dependency. It reflected hierarchical assumptions that many people accepted as inevitable, if not altogether just.

Not surprisingly, the southern colonies were the most elitist. In South Carolina and Maryland, as in Virginia, all blacks, all women, and a majority of white adult males were excluded from holding office. In these quasi-aristocratic colonies, political power was the prerogative of the privileged white males. A place on the county court was effectively limited to nominees chosen from among the wealthiest 10 percent of adult males. Many seats in the South Carolina assembly stayed in the same families for generations. In 1760, 55 percent of the assemblymen were the grandsons, sons, or brothers of former members. The white colonial gentry in the West Indies exercised similar control over local politics. In Antigua, sixty-five leading families dominated the provincial government between 1730 and 1775. Fifty of these wealthy families — those owning at least 300 acres of prime sugar-producing land — were represented in the assembly during these years, and thirty-nine were in the council.

A landed elite dominated in some of the northern colonies as well. In New Jersey, 87 percent of the assemblymen serving between 1703 and 1776 came from the richest 10 percent of the population. Moreover, 70 percent of these officeholders were members of well-established families that had settled in New Jersey in the seventeenth century; an equal percentage had one or more relatives who also served in the assembly at some point. New York also had its powerful families: the Livingstons, Van Courtlandts, and Van Rensselaers maintained an oligarchical control in the Hudson Valley, and the Delanceys dominated Westchester County.

In New England, the vast majority of adult males could (and did) participate actively in town politics. However, political leadership above the town level flowed through a limited number of bloodlines. John Adams remarked:

> Go into every village in New England and you will find that the office of justice of the peace, and even the place of representative, which has ever depended only on the freest election of the people, have generally descended from generation to generation, in three or four families at most. . . . [O]ften the disposition to favor the family is as strong in the town, county, province, or kingdom as it is in the house itself.

Adams exaggerated the degree of political exclusivity, but he did not overemphasize the importance of family ties. A single prominent family, the Williams clan, controlled county government in the Connecticut River Valley of western Massachusetts; these wealthy "River Gods" monopolized virtually all the major posts, from seats on the court to command of the militia. In Connecticut in the 1750s, the 600 offices above town level were held by only 400 men (a result of plural office holding); at any time, over half of these officials were members of the hundred most prominent families. In Hingham, Massachusetts, 70 percent of the town's selectmen in the eighteenth century were sons of selectmen. In other communities, politics became equally predictable. The great mass of ordinary men voted for the members of those families whose social prestige seemed clearly to destine them for the management of public affairs.

The Colonial Challenge to Imperial Authority

The native-born elite was merely one part of an imperial political system that spanned the Atlantic. On the highest plane, English aristocrats and bureaucrats conducted the politics of the empire. Members of Parliament, the Board of Trade, and the King's Privy Council — especially the First Lord of the Treasury and the Secretary of State for the Southern Department — all took a hand in shaping colonial policy. These London-based officials had little personal experience in America. British merchants had some knowledge of American affairs, but they imparted it largely to promote their own interests.

In the colonies, governors had the task of securing local support for imperial edicts and parliamentary legislation. Most colonial governors were Englishmen who did not derive their authority directly from the people they governed. By the beginning of the eighteenth century, only the charter colonies of Rhode Island and Connecticut elected their governors. The king appointed royal governors in Massachusetts, New Hampshire, New York, New Jersey, Virginia, and, by the 1730s, in North Carolina, South Carolina, and Georgia. The Penn and Calvert families appointed the governors in the proprietary colonies of Pennsylvania and Maryland, respectively. Therefore, governors had to draw on personal power and rely on a variety of strategies to make up for their lack of popular support. They used persuasion whenever possible, but resorted to patronage, bribery, or outright threats when necessary. Most English-born governors mobilized a few prominent colonial individuals or families to serve as their allies and agents. The more politically astute governors often played off one group of local leaders against another.

Colonial governors in the socially and politically undeveloped colonies easily managed affairs to England's advantage. In Georgia, Governor James Wright took advantage of the predominance of Scottish mercantile credit and the small population to enforce British legislation. New Hampshire's Governor Benning Wentworth, a member of a leading and long-established American family, had similar success in his sparsely settled province. He used mast and timber contracts from the Royal Navy to control the assembly in the three decades before the American Revolution. In the absence of a strong provincial elite with its own political identity and interests, a royal governor enjoyed essentially unchallenged power.

In other colonies, the emergence of a local political elite undermined the authority of the royal or proprietary governor. Local leaders developed power bases of their own and became less dependent on gubernatorial favor and patronage. They asserted their influence in the representative assemblies to increase their power at the expense of the governor's. Aggressive assemblymen used legislative control of tax revenues to limit the governor's prerogative — and sometimes even to deny him his salary.

Governor William Shirley of Massachusetts comprehended the need to accommodate the pride and power of local political leaders. In 1754, he warned his superiors in the Board of Trade that it would be unwise to disallow a rum excise passed by the legislature:

> I am persuaded that a disallowance of this act would greatly shock the minds of the representatives and disserve his majesty's government . . . I say my Lords to restrain them from exercising such a discretionary power among themselves would, I am satisfied, grieve them very much at this time, and might tend to quench that ready spirit, which they have hitherto exerted for promoting his majesty's service and the general welfare of his dominions upon the continent.

Shirley had his allies in the assembly, and he maintained an effective governing coalition during his term in office (1741–1755). He was well aware, however, that neither he nor the king's government could expect automatic acquiescence from the local legislators. The traditional weapons at the disposal of the colonial governor — a few official appointments and military contracts, his own personal prestige and political abilities — did not guarantee him sufficient power to command widespread support.

The provincial political elites often failed to consolidate their potential power, however. Instead, political factions based on family connection, economic alliance, and regional identity competed with each other for local control. In some of the northern colonies, the greater social, ethnic, and religious diversity meant that provincial politics often became extremely contentious. In New York, the Livingston family led Hudson River aristocrats and religious dissenters in a contest with the Delancey faction, which was Episcopalian in religion, mercantile in sympathy, and allied to the landowners of Westchester County. Two powerful merchant clans in Rhode Island — the Hopkins family of Providence and the Ward family of Newport — struggled fiercely for control of the powers of the elective governorship. Each faction wanted to benefit its section of the colony through the granting of monopolies, the apportionment of taxes, and the distribution of patronage. In Pennsylvania, the powerful Quaker faction gained control of the assembly in the 1720s and brought a measure of stability to Pennsylvania politics. However, the Quakers did almost nothing with their power; between 1726 and 1754, the Pennsylvania assembly passed an average of only two acts per year.

The full implications of the potential conflict between legislative and gubernatorial (and royal) power were not immediately apparent. The British government adhered to its policy of salutary neglect and made little effort to thwart the occasional assertion of legislative authority in the colonies. Even in the 1740s and 1750s, when intermittent warfare with the French required a greater British presence in North America, Britain still did not force the contest over colonial control to the crisis point. Then, in the 1760s, when Britain did attempt to assert greater control over its American colonies, it was not able to do so successfully. Both royal officials and colonial subjects suddenly confronted the limits of British authority.

The roots of this crisis lay deep in colonial society. Beginning in the 1740s, the social and political order in the colonies underwent a subtle but substantial destabilization and transformation. The very basis of authority was challenged at all levels, from the governor's mansion to the yeoman's farmhouse. This changing nature of authority in the American colonies led to dramatic, even revolutionary, consequences.

PART II

The Disintegration of the Established Order, 1740–1775

5

The Crisis of American Colonial Society, 1740–1765

During the first half of the eighteenth century, the mainland colonies grew at an amazing rate. At the time of the Glorious Revolution of 1688 in England, fewer than 250,000 settlers lived in British North America. By 1750, this population had grown to nearly 2 million. As the American population grew in size, it became more complex in its composition. The black population in the middle of the eighteenth century consisted of large but nearly equal numbers of Afro-American and African-born slaves, along with a substantial number of mulattoes. Within the white community, distinctions based on class, religion, and ethnic identity grew ever more prominent. Before the 1740s, however, these social divisions did not result in overt political strife. Then came a dramatic change. During the next twenty-five years, the American settlements experienced a series of crises that affected nearly every aspect of life.

In the 1740s, a wave of religious revivals swept over the mainland settlements. This Great Awakening increased the importance of religion in American life and changed the character of church organization and political institutions.

Then, during the 1750s, American economic life entered a new phase. The colonists increased their output of agricultural goods and imported large quantities of manufactured items from Britain. By the 1760s, many Americans resented their economic dependence on British merchants and manufacturers. The resulting increase in production and the expansion of the financial system — in the form of mortgages, store credit, and commercial loans — heightened domestic political tensions as well. Merchants and farmers struggled against one another to shape the financial policies of the colonial governments.

Moreover, a growing shortage of prime agricultural land prompted westward migration. Armed conflicts over control of western land ensued. These struggles pitted the residents of the various American colonies against one another and against the French and the Indians. A second series of confrontations, many of which were based on ethnic and religious differences, split many colonies into antagonistic eastern and western factions.

Taken together, this period of religious, military, ethnic, and economic conflict changed both the tone and the substance of colonial life. By 1765, Americans lived in a far more complicated and discordant world than they had a mere twenty-five years earlier.

The Great Awakening

An established church dominated religious life in most European countries and in many American colonies. This legally privileged institution, such as the Congregational Church in New England and the Church of England in the southern colonies, was supported by taxes paid by all residents, regardless of personal religious beliefs.

In the middle colonies, however, ethnic and religious diversity severed this traditional tie between the state and a single established church. Inspired by William Penn's vision of religious freedom, Pennsylvania's authorities eliminated state support for religion. Each sect or church had to support its own minister, using the contributions of its members. New York retained state support for religion, but allowed for religious diversity. After the English conquest in 1664, New York devised a system to allow for the multiple establishment of churches. Each community chose its own church — Dutch Reformed, Anglican, or Congregational — and supported the pastor with local taxes. Governor William Gooch of Virginia proposed a similar scheme in the 1720s and 1730s. He invited Scotch-Irish Presbyterians to settle on the Virginia frontier to create a buffer against Indian attacks. As an encouragement, he granted these Presbyterians freedom of worship and exemption from financial support of the established Church of England.

New England was not a willing participant in this evolution toward a society of sectarian diversity and religious liberty. For one thing, it received few non-English migrants. Moreover, its Puritan leaders fiercely defended their traditional principles. In the late seventeenth century, Massachusetts Bay Puritans regularly deported Quaker preachers from the colony and executed some who returned. Only in 1708 did Connecticut authorities grant Quakers, Baptists, and Anglicans the legal right to worship. And until 1727 members of these sects were taxed by the Connecticut and Massachusetts governments for the support of the established Congregational Church. Thereafter, dissenters were allowed by law to pay taxes to their own churches.

The traditional European practice of compulsory religious taxation thus remained in force except in Pennsylvania. Even in those colonies without a single established church, political leaders argued that people were too depraved to support their churches on a voluntary basis. Therefore, they used the power of the state to guarantee the financial health of religious institutions.

Events in the 1740s nearly destroyed this church-state alliance. Intense and far-reaching disagreements over religious doctrine and church practice broke out in most American sects. Religious change was particularly significant in New England. By the early eighteenth century, many New England ministers and congregations had turned away from the harsh Calvinistic beliefs of the original settlers. They no longer emphasized the innate evil of man and the overwhelming power of God. Instead, many ministers subscribed to the teachings of the seventeenth-century Dutch theologian Jacob Arminius. Arminian preachers depicted God in more loving terms. They taught that God had given men and women sufficient natural reason to attain religious knowledge and also that individuals who actively sought salvation would be rewarded by an infusion of divine grace. Because of this emphasis on human agency, Arminian doctrine as closely resembled the traditional Catholic dogma of "free will" as it did the standard Protestant belief in "free grace."

Changes in church organization accompanied this shift in doctrine. At the urging of their ministers, most congregations of Puritan Saints in New England had accepted the Halfway Covenant of 1662. This covenant allowed the children of Saints to become part-members of the church even if they had not experienced conversion. Thus, many Puritan congregations gave up the strict Calvinist doctrine of election; and they did so because of clerical pressure, a result that indicated the Saints' declining authority over religious dogma and practice.

The eminent religious historian Perry Miller referred to these changes as a "declension," a loss of the original purity and logic of Puritan practice. And so they were. But in another respect, these innovations in church practice and dogma represented the maturation of a new type of religious institution. Puritan Congregationalism was originally what church historians refer to as a sect, a breakaway group of worshippers who establish their own independent churches. It gradually evolved into what these historians call a denomination, a linked group of churches whose ministers are united in formal professional associations. Puritanism had not yet become an all-inclusive church in the traditional Catholic or Anglican sense; it was not a hierarchically organized religion, run by clergy, into which all members of the society are accepted, without having to undergo a conversion experience. But that was clearly the direction of the change.

The Great Awakening halted this trend toward an all-inclusive and less-Calvinistic church. It began as a series of local and regional revivals that had separate social and theological roots. During the 1720s, Theodore Frelinghuysen led a revival of faith and piety within the Dutch Reformed Church in New Jersey. William and Gilbert Tennant promoted similar expressions of religious zeal among New Jersey Scotch-Irish Presbyterians. And, beginning in 1734, Jonathan Edwards encouraged hundreds of young men and women in the Con-

The Reverend George Whitefield (1741). "The Multitudes of all Sects and Denominations that attended his Sermons were enormous," wrote Benjamin Franklin about this popular minister, "[and] it seem'd as if all the World were growing Religious; so that one could not walk thro' the Town in an Evening without Hearing Psalms sung in different Families of every Street." (*Source: The National Portrait Gallery, London.*)

necticut River Valley to undergo a conversion experience. Frelinghuysen and the Tennants were deeply influenced by European pietism. Pietism was an all-inclusive creed that stressed personal devotion rather than doctrinal orthodoxy as the key to salvation. Its effect in Europe had been an outpouring of religious enthusiasm among tens of thousands of peasants and artisans. Frelinghuysen and the Tennants thus preached an optimistic religious message. Because of his roots in Calvinist Puritanism, Edwards had a more pessimistic outlook. He warned his audience of a wrathful, omnipotent God who "held each sinner by a thin thread over the abyss of *hell*." Edwards called for total humility and self-abasement: "to lie low before God, as in the dust; that I might be nothing, and that God might be all."

George Whitefield, the great English evangelist, turned these distinct local revivals of piety into the Great Awakening. Arriving in Georgia in late 1739, Whitefield embarked on a series of preaching tours that carried him to nearly every colony. Whitefield's theology combined traditional and innovative doctrines. Like Edwards, he stressed that salvation came from God, through the "free gift" of grace. Yet, like his mentor John Wesley, the founder of Methodism (the English expression of pietism), Whitefield suggested that all men and women might be saved. Whitefield's great genius lay not in his theology, however, but in his oratory. A young, almost delicate man with an "angelic" countenance, Whitefield had a commanding personal presence. His words struck deep into the hearts of his audience, moving them to tears, remorse, and repentance. Thousands of people turned out to hear Whitefield speak in Philadelphia. In Connecticut, farmers dropped their plows and flocked to his sermons. Leading Boston ministers offered him the freedom of their pulpits. Whitefield's message of a wrathful, yet ultimately forgiving, God appealed to ordinary men and women. Many felt compelled to testify publicly to the power of the "New Light" that had suddenly illuminated their hearts. Others believed that God directed them to follow in Whitefield's footsteps. Within a year, dozens of self-taught preachers traveled through the colonies revealing the "Word of God" to all who would hear. The local revivals of the 1720s and 1730s merged into a truly Great Awakening.

Social and Religious Origins

Like all great outbreaks of popular emotion, the Great Awakening had diverse origins. As Jonathan Edwards reported in 1734, the initial spiritual awakening in Northampton, Massachusetts, followed "a very sudden and awful death of a young man in the bloom of his youth; who being violently seized with the pleurisy . . . died in about two days; which (together with what was preached publicly on that occasion) much affected many young people."

Then, in 1737 and 1738, a great epidemic of diphtheria swept through the colonies. Perhaps as many as 20,000 settlers, most of them children, fell victim to this "throat distemper." Like the young people of Northampton, the survivors of this epidemic were deeply impressed with the tenuous nature of life and the inevitability of death. Their feeling of utter helplessness in the face of natural forces was readily translated into religious terms. Humans seemed weak, unable to control their own destinies and in need of divine assistance.

An economic recession accentuated this feeling of dependence. In 1739, Great Britain and Spain began the War of Jenkins's Ear (so called because the Spanish allegedly mutilated a merchant skipper, Captain Jenkins, for illegal trading in the West Indies). The war disrupted the lucrative trade in molasses and farm products between the northern colonies and the sugar islands. A general European conflict, the War of the Austrian Succession, broke out in 1740. This war cut the sale of American tobacco to France and other continental nations. Together, these wars devastated the American export economy. The decline in foreign-market exports caused a drop in wholesale commodity prices of nearly 20 percent between 1740 and 1745. Unable to sell their goods abroad, Americans purchased fewer British manufactured items. Coming at the end of a two-decade period of stagnation in external trade, this wartime recession was particularly damaging. By 1745, American per capita consumption of British goods had dropped to the lowest point it would reach during the first three-quarters of the eighteenth century. (See Chart 5.2 on p. 117.)

This economic crisis crippled American port cities and commercial farming regions. The short-term recession also aggravated existing depressed conditions in hundreds of subsistence-oriented agricultural communities. As a result of a century of sustained population growth, many New England towns (and some in the middle colonies as well) lacked sufficient land to accommodate the next

CHART 5.1
Premarital Conceptions in Hingham, Massachusetts, 1650–1875

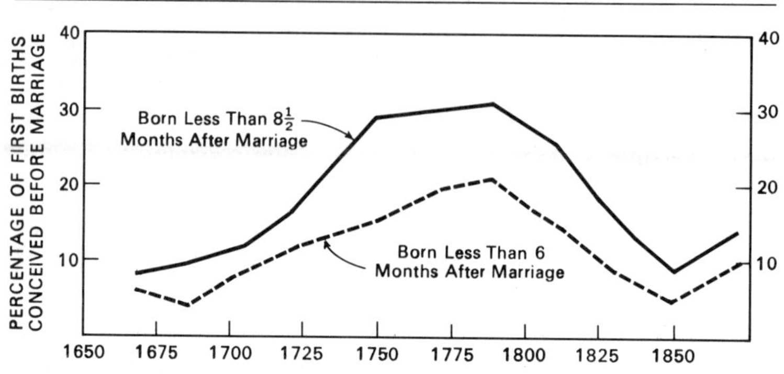

SOURCE: Unpublished data compiled by Daniel Scott Smith.

Sexual behavior is determined, in part, by historical conditions and cultural values. Between 1725 and 1825, nearly one-third of all newly married couples in Hingham had conceived a child before their marriage; before and after that period, families were generally started only after the partners had wed.

generation. This land shortage, besides creating economic hardship, undermined the traditional patriarchal order. Unable to provide farms and dowries, fathers could no longer control their children's lives.

A rapid increase in premarital sexual intercourse was one index of this crisis in parental authority. In Hingham, Massachusetts, the proportion of premarital conceptions grew from 10 percent in 1700 to 30 percent in 1750, as Chart 5.1 indicates. In Bristol, Rhode Island, one-half of the women who married between 1740 and 1760 bore a child within eight months. This alteration in sexual behavior reflected profound changes in the relations between many parents and their children. Law or custom required parental consent for marriage; in many cases, the parents "arranged" their children's marriages. Land shortages undermined this system of arranged marriages. Many parents could no longer employ economic sanctions to control their children's lives. At the same time, the evolution of a more complex agricultural economy meant that some young men and women could find employment outside the family unit, as artisans, farm laborers, or servants. Thus, young people had more freedom to choose their own sexual partners. This type of more romantic — or more defiant — union became increasingly common.

This shift in mores, like all crucial social transformations, was not accepted without struggle. Parents attempted to force their children's lives into the constrictive mold of their own experiences. Religious institutions likewise proclaimed the sanctity of the traditional morality. Headstrong children defied their parents and ministers, only to feel remorse and guilt. This widespread emotional uncertainty helped to set the stage for local revivals and the Great Awakening. In the town of Braintree, Massachusetts, in 1728, "Joseph P—— and Lydia his wife made a confession before the church . . . for the sin of fornication committed with each other before marriage." From then until 1744, the records of the Braintree congregation were filled with similar confessions of sexual transgression.

Taken together, the depressing economic, medical, and social conditions of the 1740s prepared a fertile ground — especially among the young — for the revivalists' message of human frailty and divine salvation. The "throat distemper" had killed many of their friends or siblings, and few young men and women could hope for an inheritance or for independence in a freeholding farm family. Frustrated in their cultural expectations, they sought status or solace in church membership. For example, young people figured prominently in the revival in Norton, Massachusetts. Males entered the church at an average age of 29, compared to 39 in the years before 1740. One-quarter of all new members were wage laborers, and only 36 percent owned more than thirty acres of land, compared to 79 percent of those admitted in previous decades.

The Great Awakening was much more than a response to social conditions, however. It also represented a triumph of church organization and ministerial activity. For decades, Harvard College (founded 1636) and Yale College (founded 1701) had sent scores of Congregational ministers into New England towns, thus sustaining the piety of the original Puritans through the generations. The migration of thousands of German sectarians and devout Scotch-Irish Presbyterians produced a similar set of strong religious institutions in the middle colonies. This cultural heritage of spiritual organizations and values created the necessary conditions for religious revivals.

The arrival of evangelical preachers such as Whitefield provided the crucial spark to the religious movement. Influenced by European pietism and English Methodism, these ministers energized hundreds of resident clergymen and thousands of church members. The spiritual intensity of these men and women then spread outward to infect the large "unchurched" population. During the 1740s and 1750s, the Great Awakening failed to take hold only in the Chesapeake and the Carolina colonies. Despite the efforts of the Anglican Society for the Propagation of the Gospel, these regions lacked a strong religious base on which the evangelists could build. As a result, there the Awakening would come later, in the 1760s. Whatever the timing, a combination of unsettling social conditions and purposeful human action resulted in a massive spiritual upheaval.

Religious Turmoil

The Great Awakening increased the importance of American religion while making it more pluralistic and democratic. Revivalism stressed a personal relationship with God. It appealed especially to sinners who had defied religious or social authorities by lusting after wealth, power, or sexual pleasure. Influenced by evangelical preachers, sinners confessed and humbly prostrated themselves before God. Many men and women began to look within themselves, heeding the promptings of God's grace as the ultimate arbiter of right and wrong. They denied the power of church authorities to judge their lives. Instead, revivalists convinced them to place their trust in the "Inner Light," the voice of God that spoke to them directly.

This questioning of worldly authorities threatened the power of the traditional religious and political elites. In 1740, an influential pamphlet by Gilbert Tennant warned church members of *The Dangers of an Unconverted Ministry*. The New Jersey revivalist asserted that theological training was not a sufficient qualification for clerical office; he argued that only ministers who had experienced conversion had God's authority to preach. This doctrine frightened hundreds of more conservative ministers. Their fears were well-grounded, for dozens of self-taught traveling preachers demanded the right to preach to their congregations.

The "Old Light," or conservative, clergymen banded together to uphold traditional practices and authority. They censured the revivalists, or "New Lights," as "innovators, disturbers of the peace, of the church, sowers of heresies and sedition." To buttress their privileged position, the Old Lights in Connecticut persuaded the provincial assembly to ban traveling preachers. When George Whitefield returned to the colony in 1744, he found many pulpits closed to him and his New Light followers. Legal repression failed to preserve orthodox religion and in fact undermined respect for political authority. Bitter battles shook the Connecticut assembly during the 1740s. The New Lights attempted to achieve repeal of the law against traveling preachers. They also sought relief from taxes for the support of Old Light ministers.

These struggles accentuated existing geographic and social divisions. The New Lights were strongest in the more recently settled areas of Connecticut, where much of the population was poor and in debt. These regions also demanded the issuance of more paper money (to make debt payments easier) and the expansion of settlement into Pennsylvania. Conversely, the Old Lights predominated in the older towns, where more conservative economic policies were also favored. Intense battles disrupted normal life in hundreds of towns and villages. New Lights demanded the appointment of ministers who preached with "warmth," while Old Lights defended traditional church practices and condemned "enthusiasm." Many congregations split into two, or even three, parts.

Ultimately, these divisions undermined the unity of the New Light movement in New England. First, radical separatists broke off from the moderates who sought to reform Congregationalism from within. Like the Pilgrims of the seventeenth century, these separatists demanded complete congregational autonomy and a membership strictly limited to the Saints. Then, during the 1750s, the separatists themselves divided over the issue of infant baptism. Many of them argued that baptism should only be administered to adults who had experienced conversion. They joined existing Baptist congregations or founded new churches that supported these beliefs. By the 1770s, there were more than 120 Baptist congregations in New England.

By this time, Baptists in Virginia were challenging the religious and cultural supremacy of the Anglican gentry. During the 1750s, the Anglican-controlled House of Burgesses passed strict laws to prevent frontier "missionary Presbyterians" from spreading their gospel in the tidewater region. The gentry took this action to preserve the status and power of the Church of England, which was both the legally established religion in the colony and an instrument of the gentry's authority. The gentry dominated the vestries that managed church affairs. On Sundays, their power took an even more tangible form. Gentry families arrived at the parish church in fine carriages, sat in the most prominent pews, and often publicly rebuked ministers who criticized their extravagance or moral behavior. Most freeholders, tenant farmers, and enslaved blacks encountered little more than social discrimination and formal sermons in Anglican churches. Consequently, few ordinary Virginians took an active role in the Church of England. They grudgingly paid their church taxes and avoided Sunday services.

Beginning in the 1760s, traveling Baptist preachers drew the lower classes of Chesapeake society into their congregations. The Baptists' appeal was both emotional and ideological. Their ministers preached with passion, appealing more to the hearts than to the minds of their audiences. The Baptist creed was democratic. Early Baptist leaders condemned slavery and actively sought out black converts. They encouraged church members to call each other "brother" and "sister" and to worship as equals before God. And they attacked the loose sexual code of the Virginia gentry and the preoccupation of upper-class males with drinking, gambling, and horse racing. The Baptists in Loudoun County, complained one gentleman, are "quite destroying pleasure in the Country; for they encourage ardent Prayer; strong & constant faith, & an intire Banishment of *Gaming, Dancing,* and Sabbath-Day Diversions." In opposition to the aristocratic moral code of the gentry, the Baptists proposed a puritanical and egalitarian counterculture.

The gentry used physical might as well as political power to meet the Baptist challenge. Backed by bands of armed gentrymen, sheriffs disrupted Baptist services and horsewhipped the preachers. The sheriff in Carolina County "Violently Jerked" Brother Waller off his preaching platform, a Baptist noted in his diary in 1771, and then "Caught him by the Back part of his Neck, Beat his head against the ground, [and] Give him . . . Twenty Lashes with his Horse Whip." As in the northern colonies, this repression failed. Following his whipping, "Brother Waller Went back Singing praise to God, Mounted the Stage & preached with a Great Deal of Liberty." Significantly, thousands of ordinary farmers flocked to the Baptist fold, despite the refusal of the House of Burgesses to allow them liberty of conscience or relief from taxes that supported the Church of England. On the eve of the American Revolution, one-fifth of all white Virginians had joined Baptist congregations, and Anglican ministers and gentry were on the defensive. In Virginia as in other mainland colonies, the Great Awakening transformed the character of religious life.

Westward Expansion and Armed Conflict

The sustained growth of the American population produced widespread discontent in many of the long-settled agricultural regions. The price of cleared land rose dramatically, and many farm parents no longer had sufficient acreage to provide a landed inheritance for all of their children. As fertile land became scarce and expensive, it also became the object of social conflict.

During the 1740s, a series of land riots broke out in New Jersey. These conflicts began when a small group of wealthy men revived ancestral claims to some of the best land in the province. They rested their case on proprietary grants that dated from the 1670s but had never been implemented. In fact, yeoman families had farmed this disputed land for two generations. When New Jersey courts upheld the legality of these proprietary titles, the yeomen rose in revolt, preventing the execution of legal judgments and freeing their supporters from jail. The British Board of Trade nearly dispatched royal troops to New Jersey to uphold established property rights and to suppress the riots. Not since the time of the Glorious Revolution in 1689 had events in the continental colonies seemed to require such a drastic measure.

The New Jersey land riots were not an isolated event. In most seaboard areas, there were too many children for too few farms. Some of these "surplus" youth lived at home as unmarried workers. By the 1760s, for example, 16 percent of all taxpayers in Chester County, Pennsylvania, were single freemen over the age of 21. And, in 1774, in Bristol, Rhode Island, single men and women comprised 20 percent of the adult white population.

Most young men and women who were raised on farms wanted land of their own. To achieve this goal, increasing numbers of them moved to the frontier. This migration relieved overcrowding in eastern communities, but it also generated bitter battles over western lands. During the 1750s, for example, a group of Connecticut farmers created the Susquehannah Company and laid claim to the Wyoming Valley in Pennsylvania. The company rested its legal case on the royal charter of 1662, which described Connecticut's boundaries as extending to the Pacific Ocean. Soon the company became a speculative enterprise, selling land titles to eager settlers. By 1774, more than 2000 Connecticut farmers resided in the Wyoming Valley. At the company's urging, the Connecticut Assembly formally claimed legal jurisdiction over the region.

This migration from New England challenged the land rights of the Penn family as outlined in its royal charter of 1681. The Pennsylvania legislature strongly supported the Penns' ownership of the Wyoming Valley and the entire Susquehanna River region. Soon the two colonies were on the verge of war. Settlers holding land grants from Pennsylvania trampled the fields of the Connecticut intruders and burned their houses and barns. To prevent further violence, the Penns asked the British Privy Council to resolve the dispute.

All along the frontier of overcrowded New England, similar disputes were brewing. Migrants from Connecticut and Massachusetts settled on lands claimed by the manorial lords of the Hudson River Valley. Land riots broke out on the Livingston, Van Cortlandt, and Van Rensselaer manors. Other migrants from Massachusetts and New Hampshire contested New York's control of the upper Hudson River Valley. Calling themselves the "Green Mountain Boys," these settlers used armed force to expel New York sheriffs, judges, and tax collectors.

The pattern was clear. A century of sustained population growth led to a massive migration into the interior of North America. (See map, "The Expansion of Settlement," on p. 30.) Colonial assemblies, proprietors, and landlords struggled with one another for legal control of the land, while armed settlers used force to protect their property claims.

The French and Indian War

Beginning in the late 1740s, the British Board of Trade began to realize the possible consequences of the growing size of the American settlements. Influential politicians who were serving on the board, such as Lord Halifax and Charles Townshend, appreciated the potential military importance of the populous colonies in any conflict with the powerful French monarchy. Events confirmed their judgment. In 1745, during the War of the Austrian Succession in Europe, the New England colonies attacked French Nova Scotia. Boston merchants financed this expedition to safeguard the fishing industry, which provided New England with a necessary export staple. Much to the surprise of most European observers, New England troops easily captured the French fortress of Louisbourg in Nova Scotia.

The westward movement of American settlers sparked a new series of disputes between Britain and France. Beginning in the 1740s, hundreds of migrants from the middle and southern colonies pushed across the Appalachian Mountains toward the fertile lands of the Ohio River Valley. Land speculators and colonial politicians were equally active in promoting this migration. One group of expansionists, led by Royal Governor Robert Dinwiddie of Virginia, formed the Ohio Company to seek title to some of this western land from the king. In 1754, Dinwiddie dispatched George Washington and a company of militiamen to the frontier to protect the claims of the Ohio Company from French fur traders and officials.

Dinwiddie's self-interested action led directly to the outbreak of the French and Indian War in America in 1754. When French military forces forced Washington to withdraw, the British government dispatched General Edward Braddock and 2000 regular troops to establish its sovereignty over the trans-Appalachian region. Accompanied by 1000 Virginia rangers and militiamen, Braddock advanced toward the French fortress at the junction of the Allegheny and Monongahela rivers (Pittsburgh), only to be ambushed and routed by a French and Indian force.

By 1756, these conflicts in the American wilderness resulted in the outbreak of a general war in Europe. British Secretary of State William Pitt decided

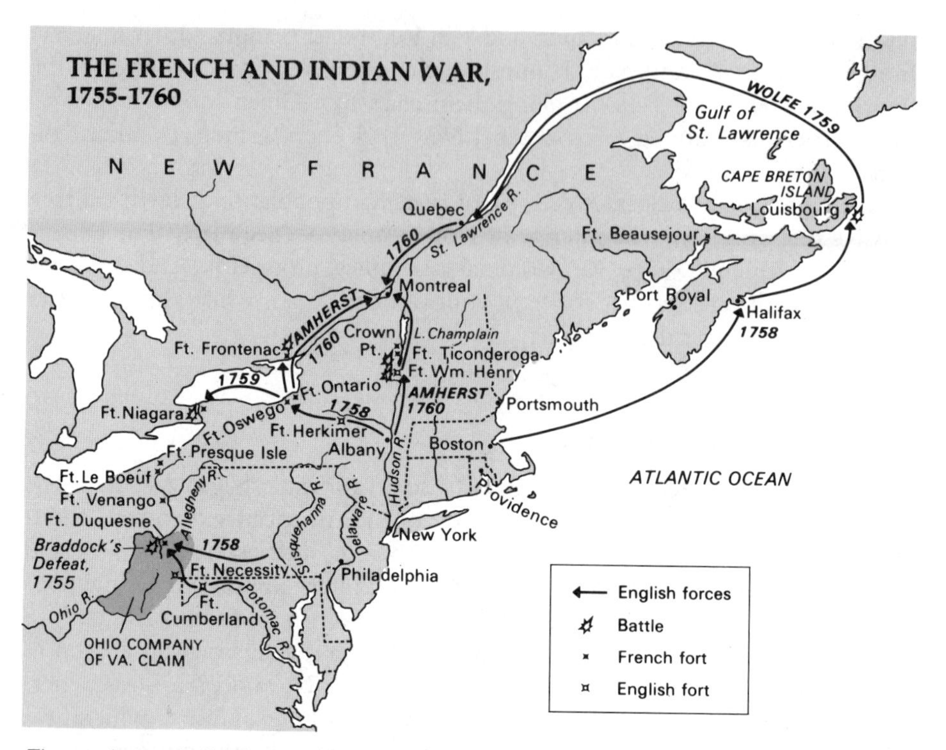

The war began in the Pennsylvania wilderness in 1755, but most fighting took place in New York and Canada between 1758 and 1760. Anglo-American forces first captured outlying French forts and then mounted combined sea and land attacks against the French strongholds of Quebec (1759) and Montreal (1760).

to confront the French not only on traditional European battlefields, but in India and America as well. Prompted by Halifax, Townshend, and other influential members of the Board of Trade, Pitt dispatched thousands of British troops to North America and paid the colonial assemblies to raise thousands of militiamen. Anglo-American forces repelled French attacks along the frontier and launched a series of overland expeditions against Canada. Warships of the Royal Navy led a military convoy up the Saint Lawrence River in 1759. Under the leadership of General James Wolfe, British troops stormed the French capital of Quebec. A year later, the British captured Montreal, effectively ending the fighting in North America. The Treaty of Paris officially ended the French and Indian War in 1763. The treaty returned two captured West Indian sugar islands to the French but awarded all of French Canada to the victorious British.

The establishment of British control over the entire North American mainland by no means ended political conflicts. In May, 1763, the Ottawa chief Pontiac led a major uprising of western tribes against the British forts that stretched from Niagara to Detroit. Fighting continued for two years and threatened to engulf the entire frontier. To quiet Indian fears, the British government issued a proclamation in the summer of 1763 that prohibited whites from settling west of the Appalachian Mountains. (See the map above.) While American speculators used their influence in London to reverse this policy, thousands of ordinary farmers migrated to the west, simply ignoring the proclamation.

In 1763, the British government also made the momentous decision to station 10,000 soldiers in the interior of North America. Three considerations underlay this decision. First, the British hoped that such a substantial force would deter both settler migration and further native American uprisings. Second, officials in London wanted to intimidate the French population in the newly conquered provinces of Nova Scotia and Canada as well as the Spanish residents of Florida, which had been captured from Spain during the war. Finally, Townshend, Halifax, and other influential ministers were determined to prevent American merchants and colonial assemblies from achieving greater autonomy. As one memorandum prepared by a senior official put it, "Troops and Fortifications will be very necessary for Great Britain to keep up in her Colonys, if she intends to secure their Dependence on Her."

The troops were used first to resolve a land dispute between rival groups of colonists, not to uphold British authority. In 1766, Royal Governor Sir Henry Moore of New York used a battalion of British troops to rout New England squatters from the estates of his political allies, the landlords of the Hudson River Valley. Newspapers in New England reported this military operation in lurid detail. Their reports of looting and indiscriminate destruction reinforced colonial fears of standing armies. This adverse political reaction prompted the British government to rebuke General Thomas Gage, the commander of the British forces in North America, for hasty intervention into a colonial dispute.

But the damage had already been done — and not merely by that one episode in New York. During the French and Indian War, thousands of British troops had been stationed in America and had left a disquieting legacy. Haughty aristocratic British officers demanded deference from wealthy colonists and political leaders and treated them as inferior and ill-mannered provincials. Average American soldiers had equally unfavorable experiences when they served with British troops. "We now see what it is to be under Martial Law," Gibson Clough, a 21-year old stonemason from Salem, Massachusetts, wrote in his diary in 1759, "the regulars . . . are but little better than slaves to their Officers; and when I get out of their [command] I shall take care how I get in again." By exposing Americans to British arrogance and military discipline, the French and Indian War awakened many colonists to some potential drawbacks of imperial rule.

The war also increased the confidence of many Americans in their own military capabilities. The colonies came out of the war with a large contingent of officers and soldiers who had actual battlefield experience, a group of merchants who were familiar with military supply problems, and many political leaders who were acquainted with the processes of mobilization and finance. As a result of their wartime experiences, a substantial number of Americans were prepared, as the imperial crisis with Britain worsened in the 1770s, to resort to war.

Economic Expansion and Crisis

Economic change also contributed to a new attitude toward Britain among the American colonists. Traditionally, commercial ties strongly bound the colonies to the mother country. British merchants handled most of the agricultural products exported by the colonists to transatlantic markets. And they dispatched increasing quantities of British manufactured goods to the mainland settlements.

Before 1755, American imports and exports were nearly equal in value. The colonists exported tobacco, rice, wheat, and fish to overseas markets. They used the funds from these exports and the income from shipping to purchase British imports. Rather suddenly, Americans went deeply into debt to their British suppliers. By the early 1760s, the colonial debt amounted to £2 million. By 1772, the American debt had risen to more than £4 million and sparked a credit crisis in Britain.

This dramatic change in the American balance of trade began in 1756, following the outbreak of the French and Indian War. As part of his worldwide military strategy for defeating the French, William Pitt dispatched thousands of troops and substantial monetary subsidies to the colonies. Military spending increased the demand for goods and services, bringing sudden prosperity to American farmers, artisans, and merchants. Wholesale prices in the colonies rose from an index figure of 69.5 in 1756 to 83.5 in 1763. This substantial price inflation stemmed partly from an increased supply of paper money. To pay for military supplies, most colonial governments issued additional currency, thus driving prices up. The war also increased the actual demand for goods, providing American producers and traders with greater profits. The colonists used their new wealth to buy imports. Between 1756 and 1760, British manufactured goods poured into America. Then, abruptly, the rapid decline in military spending brought this artificial boom to an end, as Chart 5.2 clearly indicates.

The colonies could maintain the wartime level of imports only by going into debt. British merchants were eager to extend the needed credit. The Industrial Revolution was well underway in Britain by 1750; more low-priced textiles, ceramics, and metalwares were available for sale than ever before. To dispose of these goods, British merchants gave colonial traders twelve months (rather than the traditional six months) to remit payments. By 1764, the value of British exports to the mainland colonies had again climbed to a high level. Then a postwar recession brought bankruptcy to many overextended colonial merchants. "I think we have a gloomy prospect before us," Chief Justice Allen of Pennsylvania predicted early in 1765, "as there are of late some Persons failed, who were in no way suspected, and a probability of some others, as the whisper goes."

Regardless of the problems faced by American merchants, the dominant agricultural economy remained strong. In fact, farm exports paid for a substantial portion of all imported manufactured goods. Tobacco and rice shipments increased significantly during the 1760s, and the international market price for

CHART 5.2
American Population and British Imports, 1700–1775

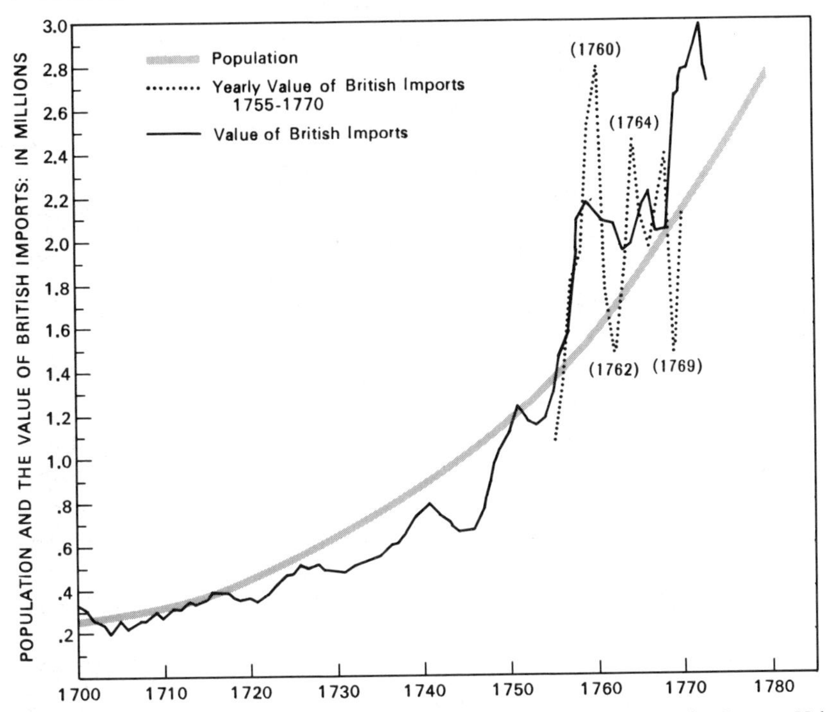

SOURCE: *Historical Statistical,* Series Z 1 and 336; John J. McCusker, "The Current Value of English Exports, 1697–1800," *WMQ* 28 (1971), Table III.

American per capita consumption of British goods remained steady between 1700 and 1720 and then declined slightly for three decades. Imports of manufactured goods soared after 1750, but fluctuated wildly in response to political events such as American boycotts.

both commodities remained high. Wheat prices shot up even more dramatically. Many European countries experienced a series of poor harvests in the late 1760s, and thus the demand for American grain increased. At the same time, England changed from a grain-exporting to a grain-importing nation as it diverted workers and resources into more profitable manufacturing activities. American farmers and merchants took full advantage of both of these opportunities. Between 1765 and 1769, the tonnage of ships clearing from Philadelphia bound for southern European ports increased threefold. The relative scarcity of grain also improved the northern colonies' terms of trade with the West Indies. After 1760, Boston merchants could barter a bushel of wheat for five gallons of molasses; two decades before, a bushel of grain brought a mere three gallons of molasses. High prices and increasing demand raised American export earnings to an all-time high.

Nonetheless, the value of imported goods rose at an even faster rate. In the five-year period centering on 1750, the average annual value of British imports to the colonies was £1.1 million. A decade later, this average annual value had soared to £2.12 million. In per capita terms, this increase was equally significant — from 19 shillings in 1750 to 26 shillings in 1760.

Due in part to imported goods, the American standard of living rose significantly. This new prosperity rested on weak foundations. It was not being completely paid for out of current earnings but was partly financed by indebtedness — by a lien on future production. Between 1768 and 1772, the value of American imports exceeded export and shipping income by more than £2 million, as Table 6.1 (p. 136) indicates. Only the money pumped into the mainland colonies by the British treasury kept this trade deficit at a manageable level. During this five-year period, Britain spent at least £1.5 million to maintain the fifteen battalions of regular troops now stationed in America. Military spending had become necessary to the stability of the American economy.

Growing prosperity and mounting indebtedness produced a great economic debate. Most Americans had traditionally consumed only what they could pay for out of current production. For them, debt was not a justifiable method of stimulating business activity. Instead, it was a moral failing that weakened personal character. John Adams, a young lawyer from the town of Braintree, Massachusetts, expressed an opinion shared by hundreds of ministers and pamphleteers. He condemned the "vicious, luxurious and effeminate Appetites, Passions, and Habits" created by the rising consumption of British manufactured goods.

By 1765, the colonies' commercial ties with Great Britain were a source of both a higher standard of living and a growing anxiety. Americans were divided within and among themselves. Even as they imported more goods, they clung to the traditional virtues of discipline, self-restraint, and frugality. Eventually, some colonists projected their diffuse economic fears outward — against the mother country. They accused British merchants and ministers of a cunning plot. By encouraging frivolous and excessive imports, these publicists argued, the British hoped to undermine the moral and financial health of the colonies and then to enslave their inhabitants by imposing a heavy burden of taxation. Thus, the end result of a decade of economic expansion was widespread suspicion of British motives, interests, and financial policy. In commercial affairs, as in military matters, the ramifications of the French and Indian War set the stage for the movement toward independence.

Ethnic Confrontations

As tensions between Britain and the colonies increased, so did political conflicts within many of the mainland settlements. By the 1760s, dissident groups had challenged the authority of many colonial governments. Social order disintegrated completely in the backcountry of South Carolina, and armed conflicts broke out in the interior regions of North Carolina and Pennsylvania. These struggles began over a variety of specific matters, such as Indian policy, the

TABLE 5.1
Ethnic Composition of the White Population in 1790

State	English	Scotch-Irish	Scottish	Irish	German	Dutch	Other
Massachusetts	84.4%	5.3%	2.7%	2.5%	0.3%	0.1%	4.7%
New York	50.3	8.7	4.3	4.1	9.1	15.9	7.6
Pennsylvania	25.8	15.1	7.6	7.1	38.0	1.3	5.1
Virginia	61.3	11.7	5.9	6.8	4.5	0.7	9.1
South Carolina	47.6	18.9	9.4	8.2	5.5	0.2	10.2
All States	59.7	10.5	5.3	5.8	8.9	3.1	6.7

SOURCE: Thomas L. Purvis, "The European Ancestry of the United States Population, 1790," *WMQ* 41 (1984), Table II.

By analyzing the names from the first United States Census (1790), historians have made estimates of ethnic composition in the late colonial period. Settlers of English descent were numerically dominant only in New England; elsewhere they were outnumbered by those of German, Scotch-Irish, and other descent (including Africans, who are not included in this table).

denial of equal political representation to settlers in western regions, and sectional discrimination in taxation. But a fundamental issue in each instance was intense antagonism among different ethnic groups.

Two long-term developments set the stage for these bitter political battles: the large-scale migration of non-English settlers to North America, and their settlement in the less crowded backcountry. In 1700, most of the white population in North America was of English ancestry. Dutch settlers in New York and New Jersey were the only significant exception. Fifty years later, English predominance was threatened by various migrant groups, especially in the southern and middle colonies. William Penn encouraged Germans to settle in his colony. He granted them separate townships, thereby facilitating the preservation of their language, customs, and religion. Other Germans settled in New York. As early as 1707, Hudson River Valley landlords welcomed shiploads of refugees from the war-torn Palatinate region, setting them up as tenant farmers. Subsequently, thousands of Germans moved down the Shenandoah Valley into the backcountry of Maryland, Virginia, and the Carolinas.

The census of 1790 revealed the extent of German migration during the first half of the eighteenth century. As Table 5.1 shows, Americans of German ancestry comprised nearly 9 percent of the total white population in 1790. In Pennsylvania, nearly 40 percent of the residents were German. Together with the Dutch, who constituted between 15 percent and 20 percent of the populations of New York and New Jersey, respectively, Germans formed a major political and cultural force in the middle colonies.

The strength of the non-English groups promoted an open and competitive political system. Ethnic and religious factions struggled with one another for political power. In Pennsylvania, the migrants threatened the traditional dominance of Quakers of English descent. As one observer noted, Baptists, Scotch-Irish Presbyterians, and German Lutherans had combined to "attempt a general confederacy of the three societies in opposition to the ruling party [of Quakers]."

The Scotch-Irish Presbyterians in the Pennsylvania coalition were the products of a complex historical experience. They were the descendants of Scottish Presbyterians who settled in Ireland in the 1640s. The English government had arranged their migration to Ireland to ensure Protestant control over the Irish Catholic majority. The transplanted Scots soon found themselves equally the victims of ethnic and religious discrimination. The English Test Act of 1704 reserved public office in Ireland for members of the Church of England, excluding Scottish Presbyterians as well as Irish Catholics. Similarly, the Trade and Navigation Acts subordinated the Scotch-Irish linen and woolen industries to those of England. Discontent grew, until finally, in the 1720s, a series of bad harvests sparked a massive migration. In July, 1728, seventeen ships left the Ulster region of Ireland for America. Over 3000 Scotch-Irish Presbyterians emigrated during that summer alone.

Many of these settlers moved to the western frontier. Their letters back to Ireland began a chain reaction of further migration. Immediate family members, more distant kin, and neighbors joined those already in America. "Read this letter," Rev. Baptist Boyd, a Scotch-Irish migrant in New York wrote back to Ireland in 1737,

> and look and tell all the poor folk of ye place that God has opened a door for their deliverance . . . for here all that a man works for is his own; and there are no revenue hounds to take it from us here . . . no one to take away yer Corn, yer potatoes, yer Lint or yer Eggs.

Between 1720 and 1776, 114,000 refugees from Ulster arrived in the mainland colonies. By 1790, the Scotch-Irish comprised 8 percent of the population of the Maine district of Massachusetts, 19 percent of the residents of South Carolina, 15 percent of the inhabitants of North Carolina and Pennsylvania, and 10 percent of the total American population.

Most migrants from Ulster bitterly resented the treatment they had received in Ireland. They refused to accept the British proclamation of 1763 that prohibited settlements west of the Appalachian Mountains. They were also unwilling to accept domination by the English inhabitants who controlled both the eastern regions of the colonies and the provincial assemblies.

The first overt conflict came in Pennsylvania in 1763. The Scotch-Irish wanted the assembly to support armed expansion into Indian territory. However, pacifistic Quakers of English descent controlled the three eastern counties, and this region elected twice as many assembly members as did the five western counties. Deprived of political redress, Scotch-Irish settlers massacred a village of peaceful Indians. Following the massacre, 250 armed men (calling themselves the "Paxton Boys") marched on Philadelphia. The city's militia, composed of tradesmen and mechanics, prepared for battle. Only the successful mediation of Benjamin Franklin, who met the Paxton Boys at Lancaster, prevented a bloody battle.

North Carolina lacked a skilled negotiator like Franklin, not to mention a tradition of Quaker pacifism. There a similar east-west conflict ended in violence. A diverse population of Germans, Irish, and Scotch-Irish lived in the Carolina backcountry along with thousands of recent migrants from Scotland (who made up 8 percent of the white population in that area by 1790). Many Scots came from the Highlands region, spoke the Gaelic language, dressed in the traditional kilt, and gave unquestioned loyalty to their clan chieftains. They had been forced from their homeland by population pressure, rising rents, and English repression after a failed rebellion in 1745. Despite their growing numbers, the Scots and other backcountry residents were under-represented in the provincial assembly, which was controlled by wealthy eastern slaveowners of English descent.

A series of local uprisings shook the North Carolina frontier in the late 1760s. Residents of Granville County and Orange County complained about corrupt elections, favoritism, and the high fees charged by provincially appointed officials. They also protested against taxes that discriminated against their less developed economy. Finally, many yeomen farmers resented the merchants and lawyers who played an increasingly prominent role in western regions. Depicting themselves as "poor families" and "industrious peasants," these farmers attacked traders and "damned lawyers who practiced numberless . . . devilish devices to rob you of your livings." Led by Hermon Husband, who condemned provincial "clerks, lawyers, or Scotch merchants . . . whose interests jar with the interest of the public good," they formed the Regulator movement. In 1768, crowds released their arrested leaders from jail, closed courts to prevent the collection of commercial debts and public taxes, and set fire to the houses and barns of unpopular judges.

Resistance by the Regulator movement continued for three years. Then, in response to desperate pleas from western merchants and public officials, Royal Governor William Tryon responded with force. Strongly backed by the provincial assembly, Tryon used the eastern militia to quell the disorder. The governor led a well-equipped army of 1200 men into the interior. He confronted 2500 poorly disciplined Regulators at the Alamance River in May, 1771. In the ensuing battle, 9 government soldiers died, and another 60 were wounded. Regulator casualties totaled 20 dead and 100 wounded. More importantly, their resistance collapsed, and 6 of their leaders were prosecuted, convicted, and put to death.

Regional tensions were equally acute in South Carolina. During and after the French and Indian War, upland settlers battled the powerful Cherokee tribe. In 1766, bands of bandits perpetrated a campaign of robbery, extortion, and arson. When the provincial government took no action to assist the predominantly Scotch-Irish and Scottish population of the region, some of these backcountry residents created another Regulator movement. This vigilante organi-

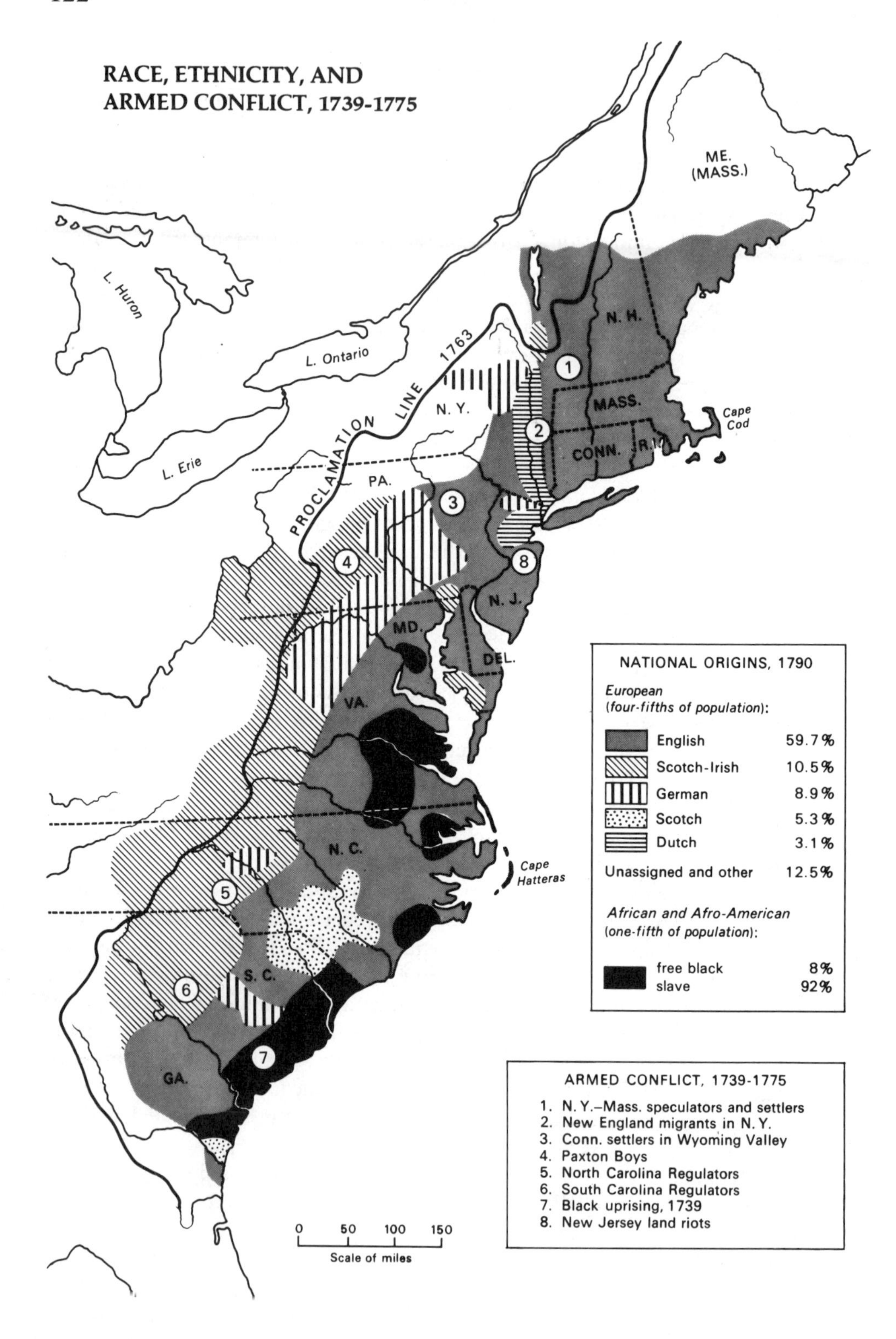
RACE, ETHNICITY, AND
ARMED CONFLICT, 1739-1775
ME.
(MASS.)
L. Huron
L. Ontario
L. Erie
PROCLAMATION LINE 1763
N. H.
MASS.
CONN.
R.I.
Cape
Cod
N. Y.
PA.
N. J.
MD.
DEL.
VA.
N. C.
Cape
Hatteras
S. C.
GA.
NATIONAL ORIGINS, 1790
European
(four-fifths of population):
English 59.7%
Scotch-Irish 10.5%
German 8.9%
Scotch 5.3%
Dutch 3.1%
Unassigned and other 12.5%
African and Afro-American
(one-fifth of population):
free black 8%
slave 92%
ARMED CONFLICT, 1739-1775
1. N.Y.–Mass. speculators and settlers
2. New England migrants in N.Y.
3. Conn. settlers in Wyoming Valley
4. Paxton Boys
5. North Carolina Regulators
6. South Carolina Regulators
7. Black uprising, 1739
8. New Jersey land riots
0 50 100 150
Scale of miles

zation had different enemies from those of the North Carolina Regulators. The South Carolina Regulators suppressed the outlaw bands. Then they began to police the "lower people," the hunters and squatters who lived at the fringes of respectable society.

Another extralegal group, the Moderation, contested the South Carolina Regulators' arbitrary assumption of power. Many Moderation leaders had previously served as provincial justices of the peace. They questioned the legitimacy of Regulator rule and demanded a return to more legalistic methods of maintaining order. In March, 1769, 600 supporters of the two rival backcountry groups exchanged shots; widespread violence seemed inevitable.

Unlike Governor Tryon, the provincial government of South Carolina would not use force to restore order. Eastern political leaders in South Carolina believed that it was too dangerous to send the militia into the western part of the colony, given the danger of slave revolts in the predominantly black lowland region. The provincial assembly therefore sought a compromise. It created four new judicial circuits in the backcountry and provided for trials by juries composed of backcountry residents. In this grant of judicial autonomy, the ethnically diverse residents of the western counties of South Carolina achieved one of the goals that had been sought by the North Carolina Regulators. Nonetheless, they still lacked equal representation in the provincial assembly and an equitable tax schedule. Like most residents of the American interior, they were consigned by the eastern elite to a position of political inferiority.

The expansion of settlement thus created new geographic divisions within most American colonies. Eastern political leaders refused to give new communities a fair share of political representation and power. One result, as the map shows, was a series of violent conflicts along the frontier. Western grievances remained strong, outlasting the defeats of the Paxton Boys and the Regulator movements. When war with Britain broke out in 1775, yeomen farmers demanded greater social equality and a more democratic political system. The people of Mecklenburg County in western North Carolina instructed their delegates to the state constitutional convention of 1776 to "oppose everything that leans to aristocracy or power in the hands of the rich and chief men exercised to the oppression of the poor."

These violent conflicts over land, taxes, and political power represented a new phase in the development of the American mainland colonies. Similar issues had often arisen in the past, but they had been resolved through the regular political process. Except on rare occasions, such as Bacon's Rebellion in Virginia in 1675 and the uprisings in Maryland, New York, and Massachusetts in 1689, white settlers had not killed one another. But the situation had changed. In New England, the tensions resulting from soil depletion and a surplus agricultural

Many armed conflicts between 1739 and 1775 stemmed from ethnic or racial tensions; others arose out of economic or cultural differences. The central issue, in each instance, was access to power — over land, labor, or political institutions.

population fueled a determined expansion into the valleys of the Hudson and Susquehanna rivers. The result was localized yet widespread violence among rival land claimants. Simultaneously, ethnic antagonisms gave sectional disputes in the middle and southern colonies a dangerous character. To those of English ancestry, the Scotch, Scotch-Irish, Irish, and Germans who migrated to North America were culturally as alien as the traditional enemy, the French. The intensity of this ethnic prejudice escalated ordinary political and economic differences into armed confrontations.

The resort to domestic violence in post-1750 America also grew out of another set of factors. These causes were racial and imperial in character. As Americans moved into the interior of the North American continent, they encountered new Indian tribes. Like the English migrants of the early seventeenth century, these settlers used force to expel native Americans from ancestral lands. The years of fighting against Indians in western Pennsylvania or South Carolina had resulted in a gradual breakdown of the prohibitions placed by civilized society on the use of force. The creation of armed slave patrols to intimidate the growing population of enslaved Africans had much the same effect within the southern colonies. Finally, New England militiamen who had killed French soldiers at Louisbourg in 1745 or in Canada during the French and Indian War had become familiar with physical violence. Members of these groups of Americans were intellectually and emotionally prepared to use weapons and force to advance individual or political goals.

Even as many Americans united together to fight the British in the War of Independence, ethnic tensions remained strong. Ethnic identity was a key factor in determining allegiance to the Loyalist or Patriot side in New York, Pennsylvania, and the Carolinas. Nor was this the only legacy of the prewar conflicts over land claims, religious liberty, and political equality. The Green Mountain Boys renounced New York's authority during the war and established the new state of Vermont. Simultaneously, Virginia Baptists bartered their support for the struggle for independence in return for legal guarantees of religious liberty. And backcountry farmers in many colonies campaigned successfully for more democratic and representative state constitutions. In these and many other ways, the internal crises that disrupted American society between 1740 and 1765 extended into the revolutionary era and shaped both its character and its outcome.

6

The Coming of the Revolution

Before 1765, the American colonists were loyal subjects of the British monarch. They had not challenged imperial authority since the time of the Glorious Revolution in 1689. For three generations, most white Americans had prospered. The Navigation Acts enacted by Parliament imposed relatively mild economic restrictions and allowed the colonists to enjoy a high standard of living. In general, the British government pursued a policy of "salutary neglect." It paid little attention to the internal affairs of the colonies, thereby permitting the development of strong local representative assemblies. The British authorities also ignored the colonists' evasion of various mercantilist regulations. The Americans used their *de facto* autonomy to create a complex and relatively self-sufficient economic system.

Then, suddenly, Britain tightened the bonds of its empire. Beginning in the 1760s, Parliament asserted its sovereignty over the American settlements and declared its intention to regulate their development. It passed laws that increased taxes in America and imposed a more rigorous administrative system. These legislative acts pursued the traditional *mercantilist* goal of subordinating the colonial economy to that of the mother country. But these initiatives were so intrusive, so numerous, and so far-reaching that they in effect created a new system of economic and political *imperialism*.

Unlike many other colonial peoples, the Americans had been prepared by their historical experience to resist this imperial coercion. Even before the first British measures were fully implemented, anti-imperial riots broke out in many colonies. The context in which these riots took place was crucial. Since the

1690s, the colonists had ruled themselves without major interference from London. Any attempt to change this established pattern, which most Americans saw as the natural state of affairs, was certain to be met with resistance. Moreover, a succession of religious struggles, economic disturbances, and armed conflicts had already unsettled the social stability of the colonies and created a fermenting political culture. Introduced into this highly charged social and political situation, the new British regulations set off an explosion that shattered the edifice of imperial authority.

The Origins of Conflict

The origins of the American War of Independence are numerous and tangled. In retrospect, it is clear that several critical preconditions for a successful movement for self-determination had been fulfilled by 1765. First, an experienced and self-confident group of political leaders had appeared in most colonies. They controlled representative institutions that enjoyed broad support among the white population. Second, three generations of rapid demographic change and economic growth had resulted in a prosperous agricultural and commercial system. Finally, changes in family life, religious practice, ethnic composition, and social authority had weakened traditional habits of political deference.

Although these social changes were prerequisites for rebellion, they were not by themselves a sufficient cause. A further set of preconditions stemmed from the French and Indian War. The war prompted an unsettling economic cycle of prosperity and recession in America. In addition, it engendered new cultural tensions. Arrogant British officers treated American "provincials" or "peasants" (as they often called them) with contempt, which led to widespread suspicion of imperial motives and policies. The war also increased the British national debt from £75 million to £133 million, setting off motions in Parliament that the prosperous colonists should shoulder part of this economic burden. Yet even these dangerously disruptive forces did not, in themselves, lead inescapably toward political violence.

The American independence movement developed finally in direct response to a third set of causes — new British measures of taxation and control. These "immediate" causes also had their origins in Britain's recent war with France. Some observant Englishmen, such as Thomas Pownall, the former governor of Massachusetts, thought the conflict with France marked a turning point in the history of Great Britain. British conquests in America and India, Pownall observed in the first edition of *The Administration of the Colonies* (1764), had created "some general idea of some revolution of events, beyond the ordinary course of things . . . something that is to be guarded against on the one hand, or that is to be carried to advantage on the other." The "revolution of events," Pownall had predicted in an earlier essay, was that "the spirit of *commerce* will become the predominant power, which will . . . rule the powers of Europe. . . . The rise and forming of the commercial dominion is what precisely constitutes the present crisis."

Once the French and Indian War had confirmed British mastery over trade with Asia, Africa, and Spanish America, the American settlements themselves would have to be brought under firm control. As Pownall put it in 1764, "Forming all these Atlantic and American possessions into one dominion, of which Great Britain should be the commercial center, to which it should be the spring of power is the *precise* duty of government."

Pownall's agenda of administrative reform reflected the sentiments of many imperial bureaucrats. Beginning in the late 1740s, the British Board of Trade recognized the growing political power of the colonial assemblies and the increasing autonomy of American merchants. At first, senior ministers ignored the Board's calls for administrative reform. Then, the French and Indian War reinforced the arguments of Pownall, Thomas Whately, William Knox, and other middle-level officials. In Massachusetts and several other colonies, the assemblies refused to raise troops unless they could appoint officers and devise military strategies — powers that the royal governors refused to concede. At the same time, colonial merchants continued to trade with French sugar islands, undermining the war effort. The British conquest of Canada and the ouster of the French confirmed the necessity of greater imperial control. British authorities realized that the colonists would no longer require military protection from Great Britain.

The imperial movement to consolidate power and enforce control began even before the end of the French and Indian War. For decades, American merchants had evaded the Molasses Act of 1733, which placed a duty of 6 pence per gallon on molasses imported from the French West Indies. They bribed customs officials to understate the size of cargoes or to pretend that the cargoes came from the British sugar islands. The Board of Trade used the war emergency to halt some of these abuses. Merchant Timothy Orme of Salem, Massachusetts, warned a ship captain in 1758:

> Since you Sailed from here our [Customs] Officers have recd Orders not to Enter foreign Molasses as heretofore. The Vessells that have arrived since those orders have been permitted to Enter about One Eighth or Tenth part of their Cargo paying 6d Sterling p. gall. Duty for what is entered — which is more than twice as much as was given before.

In 1762, Parliament passed a new revenue act that reformed the customs service. This act eliminated "absentee" officials who lived in England and received payments from easily corruptible deputies in the colonies. The ministry also directed the Royal Navy to apprehend smugglers.

This new efficiency dismayed American merchants. When customs officials in Boston used search warrants to find smuggled goods, John Hancock and other Boston merchants questioned the legality of these "Writs of Assistance" in a widely publicized trial. Speaking for the Crown, lawyer Jeremiah Gridley argued that "the subject has the privilege of house [privacy] only against his fellow subjects, and not versus the King either in matters of crime or fine." James Otis, a young and volatile American lawyer, replied that the writs were one of the "instruments of slavery." Otis maintained that the warrants were so contrary to the fundamental law embodied in the British constitution that "no acts of Parliament can establish such a writ." Merchants in Charleston, South Carolina, similarly protested against the aggressive seizure of contraband goods by customs authorities. Nevertheless, by 1765, the British customs service was collecting more than £30,000 a year in duties, far more than its £2000 annual revenue during the long period of "salutary neglect."

Other British actions had equally profound psychological and financial effects. Just as the revitalized customs service began to drain much needed specie from the American economy, Parliament enacted the Currency Act of 1764. This legislation prohibited colonial assemblies from printing paper money to use as legal tender. The colonists could no longer issue currency to provide an adequate supply of money or to stimulate business activity.

Another revenue act, the Sugar Act of 1764, likewise threatened American economic interests. The act lowered the import duty on molasses imported from the French West Indies from 6 pence to 3 pence per gallon. However, to ensure the collection of this duty, the act expanded the jurisdiction of the vice-admiralty courts. Since these courts were administrative tribunals in which judges sat alone, without juries, merchants could no longer depend on sympathetic juries of colonists to acquit them of smuggling charges. "What has America done," asked the author of one pamphlet, "to be disfranchised and stripped of so invaluable a privilege as the trial by jury?"

The British government also made other crucial changes in the American legal system. The Sugar Act transferred the burden of proof from customs officials to accused merchants, who now had to demonstrate that their trade was legal under the Navigation Acts. The Sugar Act also gave officials the authority to seize cargoes if there was the slightest "probable cause" that they were contraband, thus preempting merchants' potential suits for false arrest. Finally, the ministry instructed royal governors to issue new commissions to all colonial judges. Henceforth, judges in America would no longer hold office for life or even during "good behavior" but would serve only "at the pleasure of the crown." Lawyers in New York and South Carolina claimed this limited judicial tenure violated their "liberties and privileges" as Englishmen, but their protests were to no avail. Through a series of dramatic legal initiatives, the British Crown and Parliament had imposed new economic restrictions on the continental colonies and altered the character of their judicial systems.

Significantly, colonial newspapers gave extensive coverage to the disputes generated by the Sugar Act and to its broader implications. Before 1760, most newspaper columns were filled with news taken from London journals, and only passing notice was given to colonial events. Suddenly, many papers expanded their reporting of American affairs and encouraged readers to respond in letters to the editor. This publicity helped to make residents of the seaboard cities and many well-read planters deeply suspicious of the ultimate intentions of the British government. Aware that a dozen West Indian planters sat in Parliament,

Virginian Robert Beverley assailed the political and economic bias of the Sugar Act of 1764. It was simply unfair, Beverley charged, "that our own Interests must be totally subservient to the Luxury and Caprice of a few overgrown West Indian Planters." The New York Assembly protested that the Sugar Act was a threat to American property rights and political freedom. "An Exemption from the Burthen of ungranted, involuntary Taxes," the assembly's petition argued, "must be the grand principle of every free State." In addition, the assembly maintained, "The whole Wealth of a Country may be as effectually drawn off, by the Exaction of Duties, as by any other Tax upon their Estates." Thomas Cushing, the speaker of the Massachusetts Assembly, concluded that the act was "contrary to a fundamental Principall of our Constitution: That all Taxes ought to originate with the people."

Parliament's passage of the Stamp Act of 1765 brought these constitutional issues to the forefront. Like the Sugar Act, the new legislation was the brainchild of George Grenville, then British prime minister. Similar taxes had been collected in England since 1694; by the 1760s these taxes raised £300,000 in revenue each year. Grenville hoped to raise a similar amount in America. His legislation required the colonists to buy tax stamps from royal collectors and to affix them to a wide variety of printed materials and legal documents. Moreover, the stamps had to be paid for in sterling (rather than with colonial currency), and the law would be enforced by the vice-admiralty courts (rather than colonial common law juries).

Grenville imposed this direct tax for two reasons. First, he wanted the colonists to defray some of the costs of defending the empire. The Stamp Act specifically stated that its proceeds were to be used to pay royal officials and the British troops guarding the American frontier. Second, and perhaps more important, Grenville was determined to assert British sovereignty over the colonies and their power-hungry representative assemblies. And he wanted to do so in a way that penetrated into the very heart of colonial society. The new stamps would appear on every newspaper, book, and almanac, on every legal document and bill of lading, and on the liquor licenses and playing cards in every tavern. This symbol of imperial authority would be everywhere and would be enforced by British-controlled courts as a constant testimony to the supremacy of Parliament. Grenville's goals found a broad base of support among members of the British political elite. Despite appeals by American merchants and agents and their British associates, Parliament passed the Stamp Act by an overwhelming margin.

The Stamp Act Crisis

The Stamp Act triggered a massive wave of American resistance. On October 31, the day before the act was to go into effect, 200 New York City merchants vowed to cease the importation of British goods. Retail storekeepers and artisans promised to support this boycott and joined merchants, sailors, and laborers at a mass protest meeting. On the following night, 2000 New Yorkers surrounded the fort where British troops were guarding the tax stamps. Part of the crowd plundered the house of a British officer. Nightly demonstrations continued, and popular resistance prevented the use of the stamps. To uphold British authority, Lieutenant Governor Cadwallader Colden called on General Thomas Gage to rout the protesters by force. "Fire from the Fort," the British general replied, "might disperse the Mob, [but] it would not quell them; and the consequence would in all appearances be an Insurrection . . . the Commencement of a Civil War."

Similar violent demonstrations occurred in most commercial towns. Merchants as far away as Albany, New York, and Charleston, South Carolina, joined the nonimportation movement. A new organization, the Sons of Liberty, sprang up in most urban areas. Men from all ranks of society joined the Sons of Liberty, but most of their members were middling artisans, tradesmen, journeymen, and clerks. These men and their families made up a majority of the residents of port cities. Their livelihoods had been imperiled by the transfer of many industries — for example, shipbuilding, distilling, and sugar refining — to outlying towns and villages. In addition, many artisans in the building and leather trades faced competition from recent English and Scotch-Irish migrants. Finally, low-priced British imports threatened the jobs of colonial weavers and metalworkers. A well-established New York City resident articulated this discontent in a widely circulated pamphlet. He pointed out that America had economic interests as legitimate as those of England. If "the benefit of one must necessarily be in the same degree hurtful to the other," he argued, "then these two Interests can never unite in the same government."

Motivated by such anti-British sentiments, the Sons of Liberty used intimidation and violence to prevent the distribution of tax stamps and to force the resignation of tax collectors. In Boston, a mob burned an effigy of tax collector Andrew Oliver, destroyed his newly built office, wrecked his house and furniture, and looted his wine cellar. Twelve days later, the crowd attacked the house of Oliver's brother-in-law, Lieutenant Governor Thomas Hutchinson. To the poor laborers and middling artisans of the crowd, Hutchinson was a symbol of both British rule and upper-class wealth and arrogance. They acted accordingly. Smashing through the doors with axes, the mob knocked down all the inner walls of Hutchinson's house, leveled his beautiful formal garden, and carried off all his valuables. Only the coming of dawn prevented them from completely destroying the house. "Gentlemen of the army, who had seen towns sacked by the enemy," a Bostonian reported, "declare they never before saw an instance of such fury."

In nearly every major seaport, British authority was challenged and found lacking. Only in a few places, such as Philadelphia, was there sufficient local support to shore up the crumbling facade of imperial law and order. The ship carpenters who belonged to two organizations, the White Oaks and the Hearts of Oak, were longstanding supporters of Benjamin Franklin. Forming an "Association for the Preservation of the Peace," they prevented the Sons of Liberty from destroying the house of John Hughes, Franklin's associate and the newly appointed tax collector.

The crisis over the Stamp Act exposed the weakness of British power in America. For decades, colonial political leaders had challenged specific powers and actions of royal governors, but they had not questioned the legitimacy of the political system. Many wealthy merchants and influential artisans now directly repudiated imperial tax legislation. Along with journeymen and sailors, they joined in extralegal crowd actions aimed against British authorities and their colonial supporters.

This active opposition on the part of many American merchants and political leaders was of crucial significance. Previously the colonial elite had refused to mobilize the latent anti-British sentiments of the urban working classes. In 1747, for example, impressment gangs from the Royal Navy had swept the Boston waterfront. They seized dozens of residents for forced service in the British navy. In protest, Boston mariners, dock laborers, and artisans had rioted for three days. They burned a British rowboat and seized a naval officer as a hostage. Boston's merchants and public officials had expressed their "Abhorrence" of the crowd's action, but confined their own protest to a petition that claimed Parliament had exempted the American colonies from impressment. In 1765, many Boston merchants favored extralegal resistance and supported mob action against imperial authority. Royal Governor Francis Bernard accused forty or fifty "Gentleman actors" of fomenting the attack on Andrew Oliver's house and of instigating Samuel Adams and established tradesmen to lead the crowd.

The planters and merchants of South Carolina were equally active in achieving the resignation of the stamp tax collector. "These very numerous Assemblies of the People" that filled the streets of Charleston, Royal Governor Bull reported, "bore the appearance of common populace; Yet . . . they were animated and encouraged by some considerable Men, who stood behind the Curtain." Spurred on by Patrick Henry, a young and ardent backcountry member of the House of Burgesses, certain wealthy and influential Virginia planters decided to rouse their yeomen and tenants against the British government. "The spark of liberty is not yet extinct among our people," wrote one political leader at the height of the Stamp Act crisis, "and if properly fanned by the Gentlemen of Influence will, I make no doubt, burst out again into a flame."

Popular mobilization threatened the power of all members of the ruling elite, whether British or American. Some artisans and farmers refused to defer to traditional leaders; they wanted greater political influence for their social groups. General Thomas Gage reported that New York merchants were now "terrified at the Spirit they had raised" and worried that "popular Fury was not to be guided." Indeed, an anonymous pamphleteer in New York City advanced the radical proposition that ultimate political authority rested in the popular will. "To overthrow [the Stamp Act]," he proclaimed, "nothing is wanted but your own Resolution, for great is the Authority and Power of the People." In the end, the Stamp Act created a crisis of authority both in the British empire and in the traditional American political system.

The Roots of Rebellion

The crisis of authority precipitated by the Stamp Act had diverse roots that extended deep into the religious and political history of Great Britain and America. One source of resistance stemmed from the Great Awakening. During that time, thousands of Americans had refused to pay religious taxes and had resisted other governmental measures. Their actions undermined the traditional assumption that authorities had the right to force their views on an unwilling populace. During the 1740s, minister Ezra Stiles, a future president of Yale College, had strongly supported the political suppression of the New Lights. By 1760, however, Stiles argued that "coercive uniformity is neither necessary in politics nor in religion." By fixing the locus of divine inspiration within the heart of each believer, the New Lights also exalted the conscience of the individual, regardless of his or her social position. Isaac Backus, a New England Separatist turned Baptist, underscored the democratic political implications of the Great Awakening. "The common people," Backus proclaimed in 1768, now justly "claim as good a right to judge and act for themselves in matters of religion as civil rulers or the learned clergy."

Ultimately, religious radicalism affected the course of politics. In Connecticut, for example, both Old and New Lights opposed the stamp tax, but they did so in different ways. The Old Lights had always defended the authority of established religious institutions. Because of this conservative outlook, most Old Lights did not repudiate Parliament's authority to pass the Stamp Act; they instead petitioned the British government to repeal the tax. On the other hand, most New Lights had already challenged constituted authorities over their religious beliefs. As a result, they were psychologically prepared to take a strong stand. Connecticut New Lights joined the Sons of Liberty in great numbers and protested against the Stamp Act by participating in riots and other extralegal activities.

John Adams provided eloquent testimony as to the impact of the Great Awakening on the independence movement. As a novice lawyer in 1760, Adams heard James Otis defend John Hancock against smuggling charges. Otis attacked the Writs of Assistance used to arrest Hancock by challenging their constitutional grounds and by asserting the sovereignty of the individual's conscience. According to Adams, Otis argued

> . . . that every man, merely natural, was an independent sovereign, subject to no law, but the law written on his heart, and revealed to him by his Maker, in the constitution of his nature, and in the inspiration of his understanding and conscience. His right to his life, his liberty, no created being could rightfully contest.

Such an argument was radical in the extreme. Invoking religiously based principles of "natural law," it made the individual the ultimate judge of right and wrong. "Young as I was, and ignorant as I was," Adams commented, "I shuttered at the doctrine [Otis] taught; and I have all my life shuttered, and still shutter, at the consequences that may be drawn from such premises."

Adams was not an Old Light; his conservative outlook was based on a deep respect for legal precedents and on a secular (rather than a religious) political tradition. Adams took as given a set of assumptions and principles formulated decades before by Whig politicians in England. Most Whigs were drawn from the aristocratic and gentry classes and wanted to limit the absolute powers claimed by the Stuart kings, Charles II and James II. Their triumph came with the Glorious Revolution of 1688, which enhanced the powers of Parliament and instituted a constitutionally limited monarchy in England. The curbs imposed by the Whigs on King William III made sense to many colonial politicians. In struggles with royal governors, they copied Parliament's rules of procedure and claimed many of its legal powers. As a result, Whig principles gradually became part of the vocabulary of American politics.

During the 1730s, two other English political factions influenced the outlook of some American leaders. Many independent gentlemen who sat in Parliament espoused a "Country Party" ideology. Derived from the republican city-states of Renaissance Italy, this doctrine appealed to country gentry who feared a powerful government and were deeply suspicious of the corruption of the "Court Party" of politicians surrounding the monarch, King George II (1727–1760). Other English politicians and publicists called themselves "Real Whigs." They also attacked the power of the Court Party, which was controlled by Sir Robert Walpole. An extraordinarily clever politician and the first British prime minister, Walpole used patronage, bribes, and his close personal relationship with King George to secure firm majorities in Parliament. Real Whigs condemned Walpole's machinations as threats to genuine political debate and liberty.

The ideology of Real Whigs and Country Party adherents found an interested audience among Americans. "Bribery is so common that . . . there is not a borough in England where it is not practised," John Dickinson of Pennsylvania noted during a visit to London in 1754. Indeed, some American politicians detected a similar course of events in the colonies. They accused royal governors of using corrupt means to create subservient majorities in the representative assemblies. As early as 1732, a New York pamphleteer accused governors of seeking the "corruption and servile dependency of our . . . representatives." In 1757, the *Boston Weekly Newsletter* accused governors of "increasing the number of officers dependent on the crown and thereby influencing elections and destroying the liberties of the people."

By the time of the Stamp Act crisis, many American leaders' view of the world was deeply influenced by Real Whig or Country Party ideology. Consequently, they detected a sinister pattern in the recent British measures of taxation and control, which began shortly after the accession of King George III in 1760. Some Americans worried that the young and strong-willed monarch intended to resurrect the despotic imperial policies of James II, the last Stuart king and architect of the Dominion of New England. Others thought that Grenville was a latter-day Walpole, seeking new sources of patronage in America. If British ministers succeeded with the Stamp Act, John Dickinson warned, subsequently they would "levy upon us such sums of money as they choose to take, *without any other* LIMITATION *than their* PLEASURE." In Boston, Joseph Warren perceived an even more sinister plot. He believed that, by passing the Stamp Act, Grenville hoped "to force the colonies into a rebellion, and from thence, . . . by military power to reduce them to servitude." In any case, the new British policies were a clear and present danger to traditional colonial liberties. Even John Adams and other conservative Whigs felt the need for determined resistance.

The Significance of Economic Coercion

To oppose the Stamp Act, most colonial assemblies sent representatives to an extralegal congress in New York City in December, 1765. American political leaders, along with merchants, met to exchange information and ideas about imperial policy. The congress began a new stage in American political development. Previously each assembly had dealt on an individual basis with ministers and bureaucrats in London. Elected colonial officials had convened only once — in Albany, New York, in 1754, to discuss military cooperation against France. When the Stamp Act congress convened in December, 1765, the delegates humbly acknowledged "all due subordination" to Parliament and asked for repeal of the act. Not so humbly, they asserted that "no taxes ever have been, or can be constitutionally imposed on them, but by their respective legislatures."

American public opinion was far in advance of the sentiments of most delegates to the congress. New popular leaders, such as Samuel Adams and Patrick Henry, strongly endorsed both mob resistance and a complete boycott of all British manufactured goods. The Non-Importation Agreement subscribed to by American merchants in 1765 appealed to various colonial groups and interests. Political leaders — men familiar with the pragmatic world of power — praised the agreement's realistic approach; they felt that a successful boycott would force Parliament to respect American rights. Many Sons of Liberty also supported the agreement. As artisans who manufactured boots, sailcloth, rope, and tinware, they welcomed the boycott of the low-priced imports that competed with their products. Ministers and moralists also praised the boycott as a remedy for the disease of overconsumption that was eating away at the spiritual health of American society.

Finally, the merchant community actively supported nonimportation. Many merchants had imported huge stocks of British textiles, ceramics, and other goods in 1763 and 1764, before the postwar recession cut sales. The Non-Importation Agreement allowed them to protest against British taxes and regulations while reducing their excess inventories. "We are well convinced something of this sort is absolutely necessary at this time," Philadelphia merchant John Chew wrote to an associate in November, 1765, "from the great much too large importation that has for sometime past been made. There will be no wanted goods for a twelve month."

Economic coercion appeared to be eminently successful. Merchants in London, Bristol, and Glasgow deluged Parliament with petitions against the stamp tax. In fact, a new British government headed by Lord Rockingham had decided to repeal the Stamp Act even before news of the Non-Importation Agreement reached London. Rockingham and his associates were "Old Whigs," political descendants of Robert Walpole and the Duke of Newcastle. They stressed the commercial benefits that had accrued to Britain during the period of "salutary neglect" and maintained that a more aggressive colonial policy would endanger those benefits. Rockingham encouraged British merchants to petition Parliament for repeal of the Stamp Act, in order to counteract the arguments of Grenville and other imperialists. This political strategy also prompted the enactment of the Declaratory Act of 1766. This act asserted Parliament's power to legislate for America "in all cases whatsoever." The colonists paid a high price for the repeal of the Stamp Act, for many British politicians assumed that legislation included taxation and that the Declaratory Act defined Britain's sovereign authority in unlimited terms.

Charles Townshend certainly thought so. Townshend had served on the Board of Trade in the late 1740s under Lord Halifax, the architect of the new British imperialism, and he shared Halifax's reform goals. When Townshend became Chancellor of the Exchequer in 1767, under yet another new government, he quickly invoked the Declaratory Act to justify new taxes on America. Unlike the Stamp Act, which some colonists had condemned as an illegal "internal" tax, the Townshend Duties of 1767 raised an "external" revenue. They imposed levies on glass, lead, paint, paper, and tea imported into the colonies. Townshend reserved the revenues from these duties to pay the salaries of royal governors, thus freeing the governors from financial dependence on the American assemblies.

To resist this threatening legislation, the Sons of Liberty and the Patriot leaders once again endorsed nonimportation. Most Boston merchants supported the import ban, for it did not affect their trade with the West Indies. However, New York and Philadelphia traders refused to join the boycott, claiming that it was discriminatory. "While the importers of Wine, Molasses, etc., were pursuing their trade to considerable advantage," a Philadelphia merchant noted in 1770, "the Importers of British Goods were standing still and sacrificing all for the public good."

To enforce the controversial boycott, Patriot newspaper editors published the names of noncomplying merchants. Zealous Sons of Liberty then stoned the homes and stores of those merchants. These coercive tactics worked. American imports of British goods fell steadily. By 1769, the mainland colonies had a surplus of exports over imports amounting to £816,000, as Table 6.1 on p. 136 indicates. Acting on their own initiative, British merchant houses and manufacturers petitioned Parliament to change its taxation policies. Making what he called "a shadow Concession," Prime Minister Lord North repealed the unproductive taxes on glass, lead, and paint. At the same time, he retained the

TABLE 6.1
The American Debt to Great Britain, 1768–1772
(in £ sterling)

1768	1769	1770	1771	1772	Total
−229,000	+816,000	+87,000	−2,044,000	−826,000	−2,196,000
		British military expenditures in America			+1,500,000
			American Five-year Debt		−696,000

SOURCE: J. F. Shepard and G. M. Walton, "Estimates of Invisible Earnings in the Balance of Payments of the British North American Colonies," *JEH* 29 (1969), Table 8.

Except during the boycott of 1769 and 1770, American imports of manufactured goods far outstripped exports to Britain. Only British military expenditures in the colonies prevented a huge American debt. In a sense, the colonists paid for imperial government and military protection through their commercial purchases from Britain.

profitable tax on tea as a symbol of Parliamentary supremacy. When New York merchants accepted North's compromise, an artisan broadside condemned the action of these "Mercantile Dons" and denied the power of this "particular Class . . . to decide a Question of General Concern."

Thus, the Non-Importation Agreements of 1765 and 1768 created sharp political antagonism within America. However, these agreements also made people aware of colonial dependence on Great Britain and proposed a nationalistic alternative. "Manufacture as much as possible and say nothing," suggested Benjamin Franklin. In response, American women increased the domestic production of textiles. Crowds cheered as "Daughters of Liberty" competed in public spinning bees. In Elizabeth, New Jersey, the Heard and Woodstock families reported an output of 500 yards of woolen and linen cloth; Massachusetts and Connecticut towns turned out thousands of yards of cloth. Homespun clothes became a sign of patriotism, and the Sons and Daughters of Liberty condemned the display of imported finery.

Many politically aware Americans began to look with suspicion on all British mercantile regulations. For them, these laws had become a system of economic imperialism. They accused British merchants, manufacturers, and politicians of using their superior financial resources to exploit the relatively undeveloped colonies. One radical New York City pamphleteer condemned those Americans who protested only against direct taxation. "They have on the whole rather betrayed than defended the cause . . . ," he charged, for "tho' they condemn the Stamp-Act, [they] would have us at the mercy of the British Parliament in every article but taxation."

Even though the Non-Importation Agreements pointed the way toward a national economic policy of self-sufficiency, they did not diminish the demand for British goods. Once Parliament repealed most of the Townshend Duties, the

American balance of payments rapidly deteriorated. Between 1771 and 1774, the value of colonial imports from Great Britain was 47 percent higher than it had been between 1765 and 1768. By 1772, the Americans were once again deeply in debt to their British suppliers. (See Table 6.1.)

Economic factors, as well as Real Whig ideology and the legacy of the Great Awakening, thus played an important role in sustaining the American resistance movement. In particular, the weight of mercantile debt helps to explain the strength of the independence movement in the Chesapeake colonies. By 1776, wealthy and small-scale planters in Virginia and Maryland owed a total of over £2 million to British creditors. Many planters had gone deeply into debt to buy land and slaves and to live extravagantly. In 1776 one Virginian recalled that a debt of £1000 had once seemed excessive, but

> . . . ten times that sum is now spoke of with indifference and thought no great burthen on Estates. . . . In 1740, I don't remember to have seen such a thing as a turkey carpet in the Country except a small thing in the bedchamber. Now nothing are so common as Turkey or Wilton Carpetts, the whole furniture, Roomes, Elegant and every appearance of opulence.

Such luxury gave rise to many calls for a return to a more simple and more frugal mode of existence. Yet, as George Washington explained to a friend, his fellow planters had grown accustomed to their sumptuous style of life. Many were "ashamed" to admit that their estates could not bear the cost.

Foreign indebtedness was a new experience for most small-scale planters in the Chesapeake region. Before 1740, most freeholders or tenant farmers had borrowed money or taken goods on credit from local merchant-planters. Subsequently, Scottish merchant houses established dozens of trading stores in the Chesapeake region. To attract an ample supply of tobacco for the expanding European market, they provided funds to thousands of yeomen planters. Scottish stores spread as quickly as eager settlers did over the upland Piedmont region. In 1775, 37 Glasgow firms had 112 stores in Virginia and had extended credit to more than 32,000 planters. The 3500 debts owed to two of these firms averaged £29. Most yeomen and tenants would need two or even three good tobacco crops to repay such a substantial sum. As in the North Carolina backcountry, these small-scale planters blamed their economic hardships on Scottish merchants and the transatlantic system of British trade.

Private economic burdens added financial and emotional impact to the demand for "no taxation without representation." Most Chesapeake planters had personally experienced the anxiety of debt. Confronted by the additional threat posed by British taxes, they rose in protest. Scores of northern merchants and hundreds of urban artisans likewise felt that British administrative reforms and tax policies imperiled their financial position. Motivated both by economic interest and by ideological conviction, these Americans finally coalesced into a determined Patriot resistance movement.

Constitutional Deadlock

"What can a Governor do, without the assistance of the Governed?", a British customs official wrote plaintively from Philadelphia in 1770. "What can the Magistrates do, unless they are supported by their fellow Citizens? What can the King's Officers do, if they make themselves obnoxious to the people amongst whom they reside?" To assist local officials in upholding royal authority, the British government redeployed its American military garrisons. Beginning in 1768, it transferred some of the troops stationed in the conquered province of Canada to the seacoast cities. Unable to rule by consent, the British government resorted to the coercive means of a show of force.

To succeed, this military policy had to be combined with an astute political strategy. The government had to regain American cooperation, particularly the support of leading colonial merchants and traditional political leaders. Peace and unity were still possible, Benjamin Franklin advised the British government in 1769, if it would "*repeal* the laws, *renounce* the right, *recall* the troops, *refund* the money, and *return* to the old method of requisition."

Political instability prevented British leaders from dealing constructively with the imperial crisis. Six different prime ministers led the government in the fifteen years from 1760 to 1775. Each change in leadership brought a shift in colonial policy, as traditional mercantilists or new imperialists alternately gained the upper hand. When the American boycott of British imports prompted Parliament to debate the wisdom of the Townshend Duties in 1768, Thomas Pownall complained that

> . . . the colonies will for some time *belong to some faction* here, and be the tool of it, until they become powerful enough to hold a party for themselves, and make *some faction their tool*. . . . [There] will be one continued struggle . . . until some event shall happen that will totally break all union between us.

Pownall was only half right. He correctly discerned that shifts in British policy only fueled American resistance, but he exaggerated the significance of factionalism. The legislature was increasingly unwilling to heed opposition to its policies, from home or abroad. In 1733, Sir Robert Walpole had quickly withdrawn a domestic excise tax when mob violence threatened. Twenty years later, Prime Minister Henry Pelham also yielded to popular pressure, repealing a recently passed act for the naturalization of Jews. By 1760, however, a new generation of ministers sat in the key offices of government. They firmly believed in the supremacy of Parliament. Equally important, they adopted a rigid and legalistic approach to social and political issues. In 1766, for example, a poor harvest caused grain prices to skyrocket in many English counties. When food riots broke out, the government harshly suppressed them with military force. Two years later, the government denied a seat in Parliament to John Wilkes, a popular radical republican. The majority in Parliament continued to exclude Wilkes despite his reelection and widespread support for his cause in both England and America. When the so-called Gordon riots broke out in London in 1780, the government again acted in a repressive fashion. Troops killed 285 protesters; the mobs did not take a single soldier's or civilian's life.

This intransigent, even aggressive, Parliamentary stance hindered productive debate and achievement of compromise on difficult constitutional issues. At the time of the Stamp Act crisis, most colonial assemblies claimed the sole right to impose taxes on their constituents. In justification, they pointed both to past practice and to the venerated British constitutional principle of "no taxation without representation." Some American pamphleteers and political leaders made a finer distinction. With respect to "our EXTERNAL government we are and must be subject to the authority of the British Parliament," Richard Bland of Virginia readily admitted. "[However,] if the Parliament should impose laws upon us merely relative to our INTERNAL government, it deprives us . . . of our birthright as Englishmen." As they debated the repeal of the Stamp Act, members of Parliament forced Benjamin Franklin, the most influential colonial agent in Britain, to apply Bland's distinction to the issue of taxation. Franklin was a skilled negotiator, adept at compromise. To answer charges that Americans had rejected British authority, he replied that colonists accepted Parliament's power to levy "duties to regulate commerce" because they could always refuse to buy the taxed item. "But an internal tax," he continued, "is forced from the people without their consent, if not laid by their own representatives."

In 1767, Charles Townshend seized on this point to justify his import duties. He told the House of Commons:

> I do not know any distinction between internal and external taxes . . . if we have a right to impose one, we have the [right to impose the] other. . . . Yet since Americans were pleased to make that distinction, he was willing . . . to confine himself to regulations of Trade, by which a sufficient revenue might be raised in America.

Some colonists accepted the notion of divided sovereignty, which was implicit in this distinction between external duties and internal taxes. But most American Patriots repudiated Townshend's crafty maneuver, pointing out the true motive for his legislation. They argued that import duties designed to raise revenue (and not intended simply for the regulation of trade) were really taxes, and therefore contrary to the constitutional principle of no taxation without representation.

For American Patriots, representation was the key issue. At the time of the Sugar Act, Thomas Whately had argued that the colonists, like the majority of British people, who lacked the right to vote, enjoyed "virtual" representation in Parliament. "Every Member sits [in Parliament] . . . not as Representative of his own Constituents," the treasury official argued, "but as one of that august Assembly by which all the Commons of *Great Britain* are represented." The idea of the virtual representation of America in Parliament was so contrary to reality, Arthur Lee of Virginia replied, that it would "in the days of superstition have been called witchcraft." Lee asked:

> [Does the member of Parliament] know us, or we him? No. . . . Is he bound in duty and interest to preserve our liberty and property? No. Is he acquainted with our circumstances, situation, wants, &c.? No. What then are we to expect from him? Nothing but taxes without end.

Although persuasive, Lee's argument did not establish a convincing legal justification for the taxing authority claimed by the colonial assemblies. The British monarchy and Parliament had always claimed constitutional supremacy and had never formally granted exclusive rights of legislation or taxation to the colonial representative assemblies. No clear constitutional doctrine justified the American position. Samuel Seabury, a New York Anglican minister and a supporter of the British government, declared:

> The position that we are bound by no laws to which we have not consented either by ourselves or our representatives is a novel position unsupported by any authoritative record of the British constitution, ancient or modern.

To defend American interests on constitutional grounds would destroy the empire, Governor Thomas Hutchinson warned the Massachusetts House of Representatives, for there was "no line that can be drawn between the supreme authority of Parliament and the total independence of the colonies." Even William Pitt (Lord Chatham), an avowed "friend of America," upheld the supremacy of Parliament. While denying Parliament's right to tax the colonies, Pitt asserted "the authority of this kingdom over the colonies, to be sovereign and supreme, in every circumstance of government and legislation whatsoever."

The deadlock on constitutional issues was accompanied by the prospect of renewed violence. For decades, customs officers had ignored royal instructions because they feared hostile colonial mobs, and the Stamp Act riots of 1765 seemed to justify their taking a weak position. By 1768, however, the balance of power between the Sons of Liberty and customs officers had shifted. To uphold British authority, the ministry stationed four regiments of regular troops in Boston. The Boston town meeting strongly protested against this occupation by a "Standing Army," which, it claimed, placed them "under military rule."

The townspeople of Boston were soon at odds with the troops. Soldiers sought the company of local women and, because they were so poorly paid, competed with townsmen for part-time jobs. Residents showed their hostility by harboring deserters and by provoking the troops. Tensions came to a head in March, 1770. Soldiers fired on an unruly crowd, killing five persons. Patriot James Bowdoin published *A Short Narrative of the Horrid Massacre in Boston,* in which he accused the British of deliberately planning the killings. "Unless there is some great alteration in the state of things," Bostonian Andrew Eliot advised an English friend, "the era of the independence of the colonies is much nearer than I once thought it, or now wish it."

The presence of British troops did not restore law and order. Instead, it made clear to Americans the ultimate meaning of the new political imperialism, and it opened the way to a new dimension of conflict that threatened to develop

The Occupation of Boston, 1768. To protect customs officials from mob action, the British ministry stationed troops in Boston. "At noon on Saturday, October 1st," Patriot silversmith and engraver Paul Revere wrote, "the fourteenth and twenty-ninth Regiments . . . landed on the Long Wharf [center foreground]; there Formed and Marched with insolent Parade, Drums beating, Fifes playing, and Colours flying up King Street." (*Source: American Antiquarian Society.*)

into widespread violence. Popular opposition made the British government waver in its resolution to uphold imperial authority. The government ordered the troops to remove to an island in Boston's harbor and, for three years, avoided enacting legislation that might revive American resistance.

Toward Rebellion

In 1773, Lord North engineered passage of a Tea Act intended to assist the financially troubled East India Company. The new legislation transferred some of the company's debts to the British government and granted it various tax rebates. In addition, the Tea Act permitted the company to sell its tea directly to American consumers and retained Townshend's duty on the tea imported into the colonies. North's motivation was complex. In part, he acted to satisfy domestic political interests. More than thirty members of Parliament had close financial connections with the East India Company (compared with four members who had similar ties to the continental colonies); they demanded government assistance for the company. In part, North acted to maintain Britain's newfound position of commercial dominance; the East India Company was an important instrument of British military and economic power in India. Finally,

Political Violence in the 1770s. Some Sons of Liberty are shown protesting against the Tea Act by tarring and feathering a royal tax collector and forcing tea down his throat. The property of pro-British merchants was "actually unsafe," an observer noted, "their Signs, Doors and Windows were daub'd over in the Night time with every Kind of Filth." (*Source: Boston Public Library Print Collection.*)

like Grenville and Townshend before him, the prime minister was not content to wait for time (and gradual political pressure) to bring about American acceptance of British rule. He felt compelled to exercise Parliamentary power and to secure colonial submission. As North himself put it, "We must not suffer the least degree of disobedience to our measures to take place in that country."

Backed by Parliament and King George III, North directly challenged American merchants and the American Patriot movement. Economically, the Tea Act would hurt many American merchants, for they would no longer enjoy legal profits as wholesalers and middlemen in the tea trade. Those American merchants who illegally imported tea would suffer as well; thanks to the special tax privileges, the company's tea would be so cheap that it would drive smuggled Dutch tea off the market. By offering rock-bottom prices to American consumers, North hoped to split the American Patriot movement.

In fact, the Tea Act provoked strong and unified Patriot resistance. A group of Philadelphia merchants condemned any American who supported the new legislation as "an Enemy to his Country." Ignoring the temptation of cheap tea, most seaport artisans remained true to their constitutional principles. In Portsmouth, New Hampshire, the town meeting voted not to accept tea from the East India Company. An angry crowd forced local merchants to ship the newly arrived chests of tea to Halifax, Nova Scotia. In Philadelphia, Charleston, and other port cities, the Sons of Liberty compelled town officials to store the tea in locked warehouses. As in 1765, royal officials bowed to popular pressure.

The exception occurred in Boston. Thomas Hutchinson was at that time governor of Massachusetts and was determined to uphold imperial authority. Stamp Act rioters had destroyed his house, and members of the Massachusetts House of Representatives had stolen and published his personal correspondence. His mercantile firm, run by his two sons, was the Boston agent for the East India Company. For political, personal, and financial reasons, Hutchinson vowed to have the tea unloaded and sold. The Boston Sons of Liberty were equally determined to thwart him. A band of radical Patriots disguised themselves as Indians, boarded the ship carrying the tea, and dumped 342 chests into Boston harbor. Once again, Americans had mounted open and violent resistance to British authority.

The stakes had grown higher. The British ministry had bowed to economic coercion in 1765 and in 1768; it was no longer willing to tolerate colonial insubordination. As Chief Justice Lord Mansfield declared in Parliament, "We [are] reduced to the alternative of adopting coercive measures, or of forever relinquishing our claim of sovereignty or dominion over the colonies." Early in 1774, Parliament passed a series of Coercive Acts designed to punish the rebellious Massachusetts Patriots. The Massachusetts Government Act severely restricted the power of the House of Representatives and of local town meetings. Another act closed Boston's harbor to commerce until the East India Company was reimbursed, and yet another measure directed that Massachusetts residents could be tried for treason in other colonies or in England. The ministry was confident of the outcome. As Henry Ellis, a former governor of Georgia put it, "We know the real inability of the Americans to make any effective resistance to any coercive method which might be employed to compell their obedience. . . . What is decided upon appears judicious."

Ellis was completely mistaken. The latest legislative acts finally cut the invisible mental bonds that had tied the empire together in a peaceful political union. Some Americans who were familiar with John Locke's *Two Treatises on Government* (1690) used his theories of the social contract to repudiate British authority. They asserted that Britain had "openly dissolved the pact of society," leaving "every one to provide for his own security in the best manner he can." John Allen, a radical minister, used a different, but equally effective, line of reasoning in his influential *Oration on Liberty.* "For my part," Allen stated, "I cannot see how any Man in America, can properly break the Laws of England." "The whole lies here," he continued, drawing upon localist traditions of self-rule, "the laws of America only are broke, let the offender then be tryed by the law he has broke." According to John Adams, Allen's strongly worded tract appealed especially to "large Circles of the common People."

TABLE 6.2
British Actions and American Response, 1762–1775

1762	Revenue Act	Harrassment of customs officials
1763	Proclamation Line	Ignored by western migrants
1764	Sugar Act Currency Act	Protested in Massachusetts Discontent in Virginia
1765	Stamp Act Quartering Act	Riots and Stamp Act Congress and first Non-Importation Pact Resisted politically in New York
1766	Repeal of Stamp Act Sugar Act revised	Rejoicing in America Lower duties mostly paid
1767	Townshend Act	Second Non-Importation Pact
1770	Repeal of Townshend Act	Boom in importation of British goods
1773	Tea Act	Massive resistance
1774	Coercive Acts	First Continental Congress and third Non-Importation Pact
1775	Attack on Concord	Second Continental Congress

A strong wave of anti-British and nationalist sentiment swept across the colonies. Like most expressions of the popular will, these outbursts were not totally spontaneous. They were the final products of a decade of imperial crisis and of years of patient organization (see Table 6.2). Since 1768, Patriots had used their "Committees of Correspondence" to mobilize support in farming regions. This political agitation was about to pay off. Distressed by the terms of the Massachusetts Government Act, the colony's farmers rose in rebellion. In Worcester, militia companies ousted their old officers and appointed men who repudiated the authority of the newly appointed military governor, Thomas Gage. Armed farmers gathered on the village green, prevented the opening of the regular law courts, and forced royally appointed judges to resign their posts. The Worcester County Committees of Correspondence proposed a "convention of the people"; it would "devise proper ways and means" to create new institutions of government in Massachusetts.

Patriots in other colonies similarly repudiated British authority. In September, 1774, the gentlemen and freeholders of Fairfax County, Virginia, formed an "independent Company of Voluntiers" for "learning & practising the military Exercise & Discipline." Soon they broadened the militia training to include "all the able-bodied Freemen from eighteen to fifty Years of Age." Poorer volunteers would "form a Company of Marksmen . . . distinguishing [their] Dress . . . by painted Hunting-Shirts and Indian Boots." In Philadelphia, Scotch-Irish and German artisans took control of the city's political organizations from Quaker and Anglican merchants. They demanded support for a new pact of nonimportation to force the repeal of the Coercive Acts.

In response, politically conservative Americans also mobilized their forces. Since the Stamp Act riots, conservatives had feared the power of the mob. The mass uprisings of Massachusetts yeomen, Virginia hunters and tenant farmers, and Philadelphia artisans frightened them. They worried that financial resistance against Britain would ultimately result in colonial independence and the establishment of more democratic systems of government. In 1774, following a political meeting in New York City, the conservative aristocrat Gouverneur Morris observed:

> I stood in the balcony, and on my right hand were ranged all the people of property with some few poor dependents and on the other all the tradesmen. . . . [They] fairly contended about the future form of our government, whether it should be founded on Aristocratic or Democratic principles.

Fearing the worst, New York conservatives proposed a congress of all the mainland colonies to seek a negotiated settlement with the British government.

The First Continental Congress met in Philadelphia in the autumn of 1774. John Dickinson and Joseph Galloway of Pennsylvania rallied "men of loyal principles" behind a final attempt at reconciliation. To resolve the constitutional deadlock, Galloway proposed a new system of imperial authority. His plan called for the king to appoint a governor-general over all of the colonies and for the British Parliament to grant powers of legislation and taxation to a representative American parliament.

Dickinson and Galloway found substantial support for this proposal among delegates from the middle colonies. The diversity of social groups and political interests in Pennsylvania, New York, and New Jersey had accustomed leaders in those colonies to seeking compromise solutions. The two Pennsylvanians also sought allies among the Virginia delegation, whose members feared "the low, levelling principles" of the radical Patriots from Massachusetts. However, because of the severity of the Coercive Acts, the southern delegates were even more suspicious of British intentions. They heeded the emotional warnings of the representatives from New England. By the margin of a single vote, the delegates rejected Galloway's plan. Patriot forces in the Continental Congress then secured the passage of resolutions calling for the nonimportation of British goods and threatening to stop American exports as well. Finally, the delegates attacked the Declaratory Act and other British legislation affecting the "life, liberty, [and] property" of Americans as "unconstitutional, dangerous and destructive."

Loyalist Americans had waited too long. Most Loyalists had doubted the wisdom of the British reform movement and had initially opposed it. The situation was already desperate by the time they realized the danger the Patriot movement represented to their conservative political values. In December, 1774, New Hampshire's royal governor, Benning Wentworth, called on loyal Americans to guard the fort and weapons at Portsmouth. Only a handful of men responded. All of them were royal officials — a sheriff, two justices of the peace, four provincial councillors, and Wentworth's two personal secretaries. A month later, Wentworth's Tory Association had only fifty-nine members, and no fewer than fourteen were his relatives. Many New Hampshire residents had Patriot sympathies. Others feared retribution from the well-organized Sons of Liberty. And, as in any revolutionary situation, many men and women simply refused to become involved.

Royal authority in New Hampshire and most other colonies quickly evaporated. As they had in Worcester and Philadelphia, Patriot members of rural town meetings and county courts took control of the local institutions of government and elected their associates to the provincial assemblies. The representative assemblies, in turn, expanded their traditional powers. They passed legislation and imposed taxes without the approval of the royal governor. In most colonies, the people accepted the legitimacy of the Patriot leaders and their institutions. The social origins and ideology of the political leadership changed, but the monopoly of power remained intact.

This permanent transfer of legitimate authority did not take place without a struggle. The American political elite split into Loyalist and Patriot factions. The legal profession was a case in point. "The Lawyers are the Source from whence the Clamours have flowed in every Province," General Gage informed his superiors during the Stamp Act crisis. In fact, most older American lawyers remained loyal to the Crown. These established lawyers had grown to maturity in a legal system that emulated English practices. When New York lawyers founded a law society to debate technical issues of legal procedure, they modeled it on the moot session held by Gray's Inns of Court in London. In 1762, the chief justice of Massachusetts ordered lawyers and judges to wear English robes and gowns. In the following year, the Suffolk County bar adopted the English distinction between mere "attorneys" who handled minor legal issues and "barristers" who were admitted to plead before the supreme court. This commitment by status-conscious lawyers to English procedures and hierarchical distinctions prevented them from opposing royal authority.

Younger American lawyers became Patriots in much greater numbers than did their elders. Some of these beginning practitioners resented the fact that almost all legal and political positions were held by relatives and friends of royal officials. "Has not the Lieutenant Governor a Brother, a Judge of the Pleas in Boston?", John Adams wrote angrily in his diary in 1765, "and a Namesake and near Relation who is another Judge? . . . Has he not grasped four of the most important offices in the Province into his own Hands?" Motivated by self-interest as well as political principles, twenty-five lawyers finally signed the Declaration of Independence in 1776, along with thirty-one other Patriots.

A majority of politically active Americans had already declared their allegiance. By the end of 1774, the Patriots' control of the countryside vastly reduced the authority of the royal governors. General Thomas Gage, the newly appointed military governor of Massachusetts, actually exercised power only in Boston. In the spring of 1775, the British government ordered Gage to suppress the illegal Patriot assembly, which was meeting in nearby Concord, and to capture its leaders and its stores of ammunition and supplies. For the first time in the decade-long confrontation, the imperial authorities ordered the use of armed force. The Patriots were equally prepared for violence. Dozens of militia companies stood ready to take up arms "at a minute's notice." On April 19, these "Minutemen" met Gage's troops at Lexington and Concord. The Patriot soldiers inflicted heavy casualties on the British troops as they retreated back to Boston.

In the ten years since the Stamp Act crisis, the colonists had moved from verbal resistance to armed rebellion. At first, Americans had simply defended their traditional privileges within the British Empire. The debate over taxation led them onto new constitutional and political ground. American political leaders drew upon the Whig and Country Party ideology to justify their opposition to the new policies of economic and political imperialism. The force of events prompted the colonists to define their self-interests as members of distinct social groups and as residents of America. By 1775, the possibility of reconciliation with Great Britain was slim.

In addition, the authority of the elite, the traditional leaders of American society, was fragile. During the Stamp Act riots and the nonimportation movements, urban artisans and freeholding farmers demanded government based on popular sovereignty and political equality. This making of "sovereigns [out] of subjects," as the New York Patriot John Jay put it, represented a new stage in the long-term evolution of American society. The Great Awakening had aroused latent republican (and democratic) tendencies; the imperial crisis made them manifest. "All of the miseries of mankind had arisen from freemen not maintaining and exercising their own sentiments," a Philadelphia artisan would argue in 1784. "No reason can be given why a free people should not be equally *independent* in . . . their *political* as well as their religious persuasions." The disintegration of traditional institutions and values that had begun in 1740 was now complete. With the repudiation of British rule and the end of deference to their traditional rulers, the American people had entered a new political era.

PART III

The Creation of a New Institutional and Social System, 1775–1820

7

Military Mobilization and Revolutionary War

John Adams once described the outbreak of the American Revolution as thirteen clocks striking at once. With an upsurge of patriotic fervor, people from all the colonies proclaimed their commitment to defending their rights. After taking up arms in 1775, breaking away from England was the logical, almost inevitable, next step. When rebellious Americans declared their independence in 1776, they further defined their collective identity. They spoke of themselves not just as the inhabitants of thirteen separate states, but as a single people united in the common cause of liberty.

The emphasis on common cause masked deep divisions, however. Adams also observed that only a third of the American people actually supported the Revolution, a third remained loyal to England, and another third were largely indifferent. Adams's estimate of the proportions may have been off, but his perception of essential divisions was correct. Americans did not universally favor the independence movement. Even supporters of the Revolution did not agree about its goals. Eight long years of war made the flaws in American unity increasingly obvious. Many people pursued regional, social, and personal goals, sometimes at the expense of their commitment to the revolutionary struggle. Profiteering, mutiny, and treason became almost as common a part of the revolutionary experience as sacrifice, duty, and patriotism. By the end of the war in 1783, Americans had achieved a considerable, but costly, victory; they had gained their independence, but they had also lost much of their initial idealism.

Volunteer Revolutionaries and Republican Virtues

The news of the fighting at Lexington and Concord on April 19, 1775 jolted New England Patriots into action. In nearly all communities, companies of Minutemen assembled hastily and, usually in a matter of hours, set off for Boston. Within a few days, more than 10,000 citizen-soldiers were encamped outside Boston, determined to "show the whole world that a Freeman contending for his LIBERTY on his own ground is superior to any slavish mercenary on earth." Most militiamen volunteered to stay with the Patriot army for eight months — long enough, they thought, to achieve victory.

Throughout the spring of 1775, American troops were remarkably successful. The main body of the New England militia kept the British troops bottled up in Boston. Meanwhile, a smaller force captured Fort Ticonderoga on Lake Champlain in May. Completely surprised by the American attack, the British defenders surrendered the fort and, even more important, relinquished much-needed artillery. A month later, the British attempted to break the siege of Boston by attacking American emplacements on Breed's Hill in Charlestown. After three almost suicidal assaults on the entrenched Americans, the British troops took the hill. The engagement (mistakenly called the Battle of Bunker Hill) was an extremely costly one: the British suffered over 1000 casualties out of a force of 2500, more than they would lose in any other battle of the war. As the shaken British general Henry Clinton lamented, it was a "dear bought victory . . . another such would have ruined us."

The British were understandably demoralized by the success of the amateur American army against a professional European force. General Thomas Gage admitted that he had greatly underestimated the will and fighting skill of the American militia units. "These People Show a Spirit and Conduct against us they never showed against the French," he wrote, "and every body had Judged of them from their former Appearance, and behavior . . . which has led many into great mistakes." For the next ten months, the British avoided major offensive actions.

While the British remained in defensive positions, the Americans took the opportunity to strengthen their forces. In June, 1775, the New England militia companies asked the Continental Congress for official recognition and reinforcements. The congress immediately dispatched militia units from Pennsylvania, Maryland, and Virginia. More important, the congress sent George Washington to take command of the army. Washington clearly wanted the job — he wore his Virginia militia uniform to the sessions of congress — and the congress wanted Washington. As John Adams argued, having a southern general conduct a northern campaign would underscore the common commitment of the colonies.

When Washington arrived in Massachusetts to take command officially on July 2, 1775, he faced the difficult and discouraging task of turning a collection of militia units into an effective fighting force. Like the British commander

"The Yankee Doodles Intrenchments." One soldier (the Indian recruit at the far left) confesses his lack of courage; another (second from the right) refers to the lowly occupational status of the American commander, General Putnam. (*Source: The British Museum.*)

Gage, Washington had served in the French and Indian War. He knew from experience that the colonial militia was often unreliable, especially in a protracted campaign. Militiamen were accustomed to fighting close to home for short periods of time under officers they had elected from their own communities. Because of this localist orientation, they did not adapt easily to being part of a standing army with a professional officer corps. Indeed, many American rebels, both in the army and in the Continental Congress, were uneasy about creating a standing army. Professional troops had associations with the tyranny so hated in the British; many Americans preferred to rely on the republican virtue and spirit of a volunteer army.

By the fall of 1775, many of the New England militiamen had lost their initial enthusiasm and refused to reenlist. Washington complained disgustedly about this lack of republican resolve: "Such a dearth of public spirit, and want of public virtue, such stock-jobbing . . . I never saw before and pray God I may never witness to again." He knew that short-term volunteers could not ultimately defeat a well-trained, well-equipped professional army. Given the widespread belief in the virtues of voluntarism, however, Washington dared not press too hard for military conscription or long enlistments. Instead, he instituted and enforced rules of military discipline on the troops under his command in Boston. He subjected all his men to regular training; those who disobeyed orders were flogged.

By the spring of 1776, Washington's army had become strong enough to dislodge the British from their positions. When the Americans placed the cannons captured at Fort Ticonderoga on hills within range of Boston, the British decided to withdraw. On March 17, British troops sailed for Nova Scotia, accompanied by about 1000 Loyalists.

British Assets and Allies

The British evacuation of Boston was not a surrender, but a strategic shift. The British still enjoyed military superiority, and they controlled the sea. While the naval blockade of American ports limited the flow of foreign supplies to Washington's army, British officers received ample supplies and reinforcements from England.

Moreover, the British had considerable support among American Loyalists, especially in the middle and the southern colonies. Some Americans — government officials appointed by the king, merchants connected to British firms, clergymen and laypeople in the Anglican church — remained loyal to Britain because of strong ties to British institutions. Others sided with the British because their political enemies were on the Patriot side. Many tenants in the Hudson Valley, for instance, detested the Livingstons, Van Rensselaers, Schuylers, and other elite landholding families, whose members emerged as leaders of New York's revolutionary movement. Similarly, former Regulators in North Carolina refused to join forces in 1775 with the political leaders from the eastern part of the colony who had sent troops against them in 1771. The proportion of active

Loyalists never quite reached the third of the population suggested by John Adams; the estimated 500,000 Loyalists represented only about a fifth of the population. Still, they played a crucial role in the British war effort by providing soldiers, supplies, and spies.

The British also took advantage of racial divisions in the South. In November, 1775, Lord Dunmore, the royal governor of Virginia, issued a proclamation granting freedom to any slaves who would take up arms against the forces of the Patriot movement. Dunmore was not interested in freeing black slaves — he was a slaveowner himself. What he hoped to do was to create anxiety among the white Patriots and, if necessary, to use blacks to suppress the rebellion. To many slaves, the prospects for freedom looked better with the British than they did with the Americans. Almost 1000 blacks fled from their masters to join Dunmore, who organized several hundred of them into a military unit. Wearing the slogan "Liberty to Slaves," Dunmore's "Ethiopian Regiment" joined British regulars in battle against white Virginians in December, 1775. Dunmore's force lost, however, and he retreated to British ships cruising off the coasts of Virginia and Maryland.

In the long run, neither Dunmore's Ethiopian Regiment nor his proclamation had a significant bearing on the military outcome of the war. The slaves who escaped to Dunmore's ships found no safe haven, but a death trap. Disease swept through the ships in the spring of 1776, and the majority of blacks on board died. Still, slaves continually fled to the British lines throughout the war years in the hope of gaining their freedom. Others took advantage of the confusion of wartime to escape from their masters and pose as free blacks in American society. Contemporary observers estimated the number of runaway slaves in Virginia and South Carolina alone at more than 50,000. As Dunmore had hoped, the slave population remained a source of uneasiness for white revolutionaries in the South.

From Rebellion to Revolution

Throughout the first year of the war, American rebels did not openly seek total independence from England. Many Patriots still considered themselves loyal subjects of the king. They maintained that the problem lay with a corrupt Parliament, which had imposed tyrannical laws and taxes and then sent a standing army to enforce them. These rebellious Americans resorted to armed force only to protect, or regain, their just rights as English citizens.

Seeking conciliation, the Second Continental Congress drafted two appeals to the king in July, 1775: the Declaration of the Causes and Necessity of Taking Up Arms, and the Olive Branch Petition. Both documents argued that the Americans were acting only in defense of their rights. The Declaration of Causes stated:

> . . . we do not mean to dissolve that union which has so long and happily subsisted between us, and which we sincerely wish to see restored. — Neces-

> sity has not yet driven us into that desperate measure, or induced us to excite any other nation to war against them. — We have not raised armies with ambitious designs of separating from Great Britain, and establishing independent states. We fight not for glory or for conquest.

King George III ignored these appeals. In August, 1775, he declared the colonies to be in open rebellion and ordered a major military mobilization. Parliament soon followed suit; it ordered the seizure of American ships and the end to all trade with the colonies. The British government had clearly chosen retribution over reconciliation.

American radicals no longer hesitated to demand a total break from Britain. In January, 1776, Thomas Paine forcefully articulated the radical position in his best-selling pamphlet *Common Sense*. Paine attacked the very institution of monarchy, especially the idea of hereditary rule. William the Conqueror, he sneered, was a "French bastard landing with armed banditti and establishing himself king of England against the consent of the natives." George III was no better — a "Royal Brute." When the king's troops fired on American yeomen at Lexington, Paine wrote:

> I rejected the hardened, sullen-tempered Pharoah of England for ever; and disdain the wretch, that with the pretended title of FATHER OF HIS PEOPLE can unfeelingly hear of their slaughter, and composedly sleep with their blood upon his soul.

The time for petition had passed, Paine concluded:

> Every quiet method for peace hath been ineffectual. Our prayers have been rejected with disdain; and hath tended to convince us that nothing flatters vanity or confirms obstinacy in kings more than repeated petitioning. . . . Wherefore, since nothing but blows will do, for God's sake let us come to a final separation, and not leave the next generation to be cutting throats under the violated unmeaning names of parent and child. . . . A government of our own is our natural right.

Paine's call for independence was echoed throughout the colonies. In June, Richard Henry Lee put before the Continental Congress a resolution for independence proposed by the Virginia provincial convention. The congress appointed a committee of five — John Adams of Massachusetts, Roger Sherman of Connecticut, Robert Livingston of New York, Benjamin Franklin of Pennsylvania, and Thomas Jefferson of Virginia — to compose a statement of American goals. Jefferson drafted the Declaration of Independence, and after making several revisions and deletions, Congress adopted it on July 4.

Like Paine, Jefferson no longer distinguished between a benign king and a corrupt Parliament. "The history of the present King of Great Britain is a history of repeated injuries and usurpations," he wrote. Jefferson went on to present a long list of specific (if somewhat exaggerated) grievances. He argued that these reprehensible actions were not the result of incompetent administration, but rather the product of a malicious and conscious design that had "in direct object the establishment of an absolute Tyranny over these States." In the face of this evil plan, Americans had no choice but to fight for their independence: "it is their right, it is their duty, to throw off such Government, and to provide new Guards for their future security."

The Declaration of Independence contained no strikingly original ideas. Most of its premises — and even some of its phrases — were taken from the writings of the seventeenth-century English philosopher John Locke or the eighteenth-century philosophers of the Scottish Enlightenment. Its emphasis on corruption and conspiracy in the British government also reflected the Real Whig ideology so often articulated following the Stamp Act crisis of 1765. Jefferson's fellow delegates already knew and accepted the basic argument of the Declaration of Independence. As John Adams wrote, "there is not an idea in it but what had been hackneyed in Congress for two years before." But the Declaration did set forth the rationale for revolution eloquently as well as emphatically. The fine words had little effect on British officials, but they gave an enormous boost to the spirits of the American people.

The War in the North, 1776–1778

Unfortunately for the revolutionaries, the military power of Washington's army did not equal the rhetorical force of Jefferson's words. After the British evacuation of Boston, Washington moved his base of operations to New York City, where he split his troops between Long Island and Manhattan. The British also considered New York to be a critical prize because of its strategic location and excellent port. Under the command of General William Howe and his brother, Admiral Lord Richard Howe, a huge combined force of over 30,000 British soldiers and sailors was sent to take New York. In August, 1776, this force attacked Washington's troops on Long Island and inflicted one of the worst defeats of the war on the Americans. Had the Howe brothers followed up this victory with a quick move against Washington's forces in Manhattan, they could have smashed his army and perhaps won the war. Their indecisiveness allowed Washington to escape to New Jersey and then to Pennsylvania.

There Washington took a small measure of revenge in two bold attacks on the British forces who had pursued him through New Jersey. He surprised a sleeping Hessian encampment at Trenton on Christmas Eve, 1776; ten days later, his troops defeated a column of British regulars at Princeton. Neither of these victories had important strategic value, but they helped to bolster American morale.

The most important American victory of the early part of the Revolution, and perhaps of the whole war, came at Saratoga, New York, in September, 1777. There soldiers of the Continental Army and New England militia units led by General Horatio Gates surrounded a large British force of almost 10,000 men. The British general, "Gentleman Johnny" Burgoyne, was advancing down the Hudson Valley to join forces with General Howe's troops in New York City.

THE WAR IN THE NORTH 1775-1777

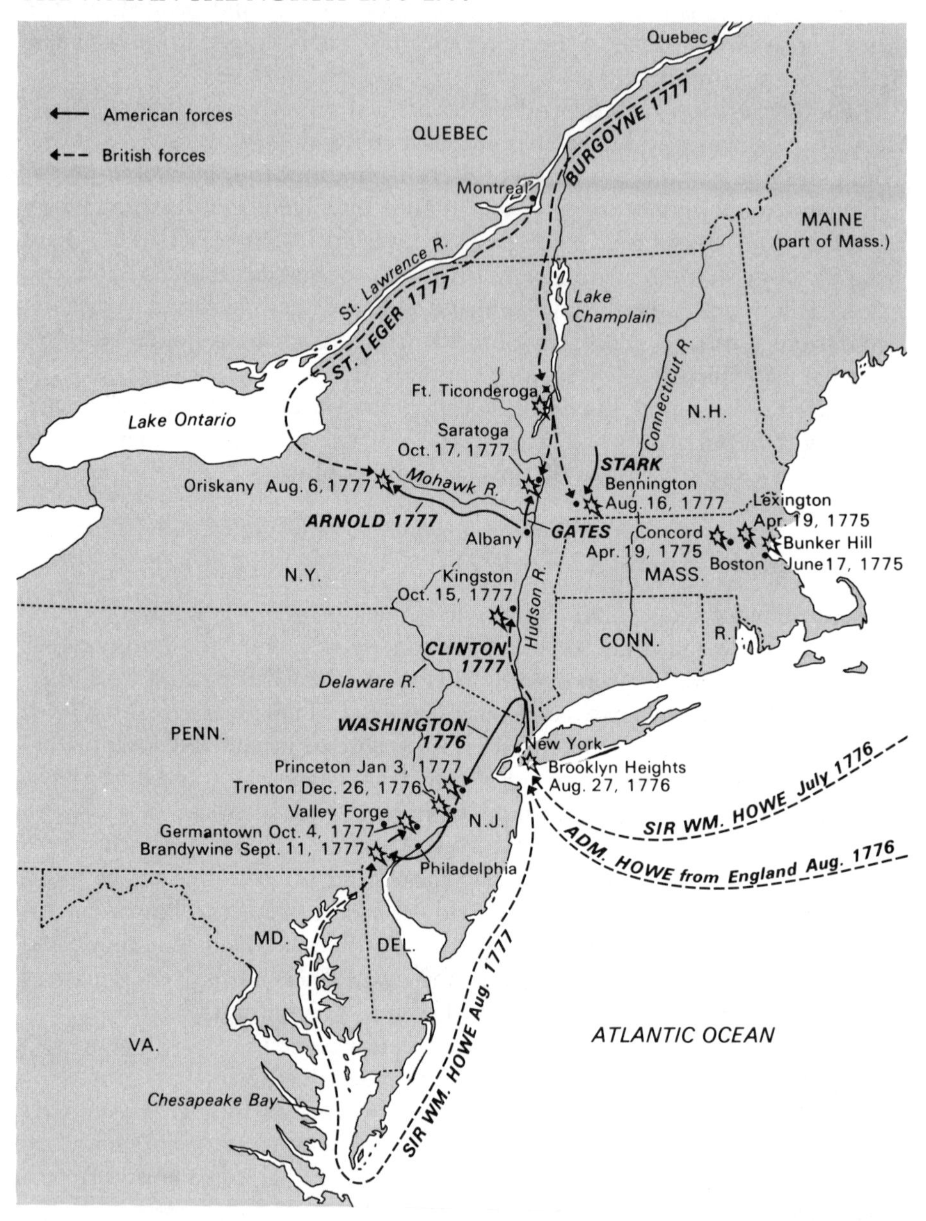

By the end of 1775, Patriot forces controlled most of New England. The critical American victory at Saratoga in 1777 allowed New England to continue support to hard-pressed Patriot troops in British-dominated New York, New Jersey, and Pennsylvania. By 1778, the war in the North had essentially reached a stalemate.

Gates — aided by a brilliant young officer under his command, Benedict Arnold — defeated the slow-moving Burgoyne in two engagements and forced him to surrender. Burgoyne lost more than his army — as bad as that was; he also lost the opportunity to cut New England off from the rest of the colonies.

On their part, the Americans won far more at Saratoga than the battle. Their impressive victory paved the way for the French to enter the war as their allies. A longstanding enemy of Britain, France had been providing the Americans with covert assistance since 1775, in the form of money, munitions, and supplies. American commissioners in France, led by Benjamin Franklin, had been pushing for an open alliance. The French government had consistently refused, fearing the imminent collapse of the American rebellion. Reassured by the victory at Saratoga, the French finally signed a formal treaty of alliance in February, 1778. With the entry of France into the war on the American side, Britain could no longer be so confident of its control of the sea and even had to worry about attacks from across the English Channel. The Americans, conversely, now had ready access to European supplies and soldiers.

The Time of Trial at Valley Forge

The French alliance did not come soon enough to help Washington's army through the winter of 1777–1778. The war in the North essentially froze to a halt. In September, 1777, Howe's British forces had attacked and occupied Philadelphia, where they enjoyed the comforts of America's largest and wealthiest city. Less than twenty miles away at Valley Forge, Washington's troops established winter quarters in the open countryside. Military success had to be postponed; their goal was mere survival. The grim winter was a severe test for the men and women at Valley Forge, and they showed both the best and the worst sides of the American revolutionary.

The biting cold and brutal camp conditions devastated Washington's army. Of the 11,000 Patriots who marched into Valley Forge in December, over 2500 were dead by spring — not from hostile military action, but from malnutrition and disease. Ill-clothed, underfed, and crowded into hastily built log huts, the American troops brooded and grumbled about their situation. When food supplies ran low, they gave voice to a menacing chant, "No bread, no soldier." Some made good on this threat; desertions averaged about ten a day. The rest of the men, with a sizable number of women camp followers, made do as best they could. They cooked and cared for each other, foraged for food and fuel, and tried to stay warm in tattered shoes and coats.

The civilian population in the surrounding countryside provided little aid to the American troops. To avoid damaging civilian morale, Washington did not publicize the sad situation existing at Valley Forge. Some civilians therefore had no idea what conditions in the camp were like. Among those who did know, many put profit before patriotism. They refused to sell provisions to Washington's army because they preferred to deal with the British, who paid in gold and silver rather than rapidly depreciating currency issued by the Continental Congress. The American army commandeered supplies when necessary, and some soldiers stole from civilians whenever possible.

The Valley Forge experience was a sign of the growing separation between American soldiers and citizens. In the three years since the battles at Lexington and Concord, patriotic enthusiasm had faded, as had the notion of a volunteer army of citizen-soldiers motivated by republican virtue. In late 1776, the Continental Congress reluctantly agreed to Washington's repeated requests for a long-term standing army. It established manpower quotas for each of the states and instituted military conscription. As an inducement to volunteers, the congress also offered bounties to men who would enlist for the duration of the war. By 1778, most soldiers in the Continental Army were conscripts or long-term enlistees. Also, most of them were young, propertyless, and poor. Fully 90 percent of New Jersey's Continentals came from the lower two-thirds of the economic ladder; almost half of them had no property at all. Wealthier men easily avoided service in the Continental Army by sending hired substitutes or, in some cases, black slaves in their places. The social composition of Washington's army became very different from that of American society in general. Here was another reason for civilian indifference to the plight of the army at Valley Forge: in war as well as in peace, the poor are often pushed out of sight and forgotten.

Many Continental officers, by contrast, came from the upper levels of American society. They had been leaders of their communities in peacetime, and they looked upon military service in wartime as the proper responsibility of a gentleman. Accordingly, they expected remuneration and respect commensurate with their social position. However, most of these officers had to endure the same privations as the enlisted men. Keeping up gentlemanly appearances under such trying conditions was expensive, and an officer's pay was meager. Most officers had to spend their own money for proper uniforms, equipment, and food. Some, like General Anthony Wayne, also used their private means to purchase clothing and supplies for their men. As officers depleted their personal resources, they saw civilian merchants and agents amassing sizable fortunes through dealing in military supplies.

The contrast between their own idealistic sacrifice and the civilians' opportunistic success created bitterness among the members of the officer corps. In November, 1777, a group of officers asked Washington for his assistance. They wanted the Continental Congress to guarantee them a half-pay pension for life. A pension would restore some of their money and recognize the service they had rendered. Washington knew that the congress would not be easily convinced. The president of the congress, Henry Laurens of South Carolina, looked upon the demand for pensions as "a total loss of virtue in the Army." Washington knew that the stakes were high. If the congress refused, many officers would resign in protest. He reluctantly threw his weight behind the pension proposal. As he wrote to Laurens in April, 1778, "I do most religiously believe the salvation of the cause depends upon it." The congress finally agreed to fund officers' pensions — but only for seven years, not for life. The pension controversy further highlighted the growing sense of enclosed identity within the Continental Army. Officers considered themselves an elite brotherhood of gentlemen-soldiers, with professional honor and interests to protect. They began to regard civilians, in both the Continental Congress and society at large, as their political and social adversaries.

Throughout the winter at Valley Forge, nothing had a more remarkable effect on the professionalism and pride of the Continental Army than the arrival of a former Prussian army officer, Friedrich Wilhelm August Heinrich Ferdinand, Baron von Steuben. "The Baron" served as a volunteer training officer at Valley Forge, and he put great emphasis on repeated and rigorous drill. Even though he spoke little English, he communicated to Washington's soldiers the importance of discipline. His military-manual methods imposed greater order on the way American soldiers marched and fought. Just as important, his concern for the common soldier imbued the troops with increased self-respect.

When Washington's troops greeted the spring of 1778, they had endured not only a hard winter, but also hard training. Both experiences gave them a new sense of accomplishment and confidence. Unable to count on civilian support, the officers and enlisted men at Valley Forge had found the necessary material and spiritual resources on their own.

War on the Home Front

The war tested civilians as well as soldiers. Few civilians faced rigorous discipline or violent death, but many experienced major disruptions in the pattern of normal life. During the course of the war, people had to find new ways to govern their communities and provide for their families. The economic and psychological hardships of war mounted with each year; by 1783, the common people had undeniably endured their share of sacrifice and suffering.

Even before armed conflict broke out, civilians faced some hard choices. As early as 1774, normal governmental activity ceased. In reaction to the Coercive Acts, angry people shut down many of the basic institutions of government. They repudiated the authority of Loyalist officials and subjected them to humiliation and abuse. In Hampshire County, Massachusetts, a mob of local farmers descended on Israel Williams, the most powerful political figure in western Massachusetts and a leading Loyalist. They shut Williams and his son up in a cabin, stuffed the chimney, and smoked them overnight. In the eyes of some observers, this outburst of extralegal activity threatened to bring about social and political chaos. One Massachusetts minister fretted, "Every Body submitted to our Sovereign Lord the Mob — Now we are reduced to a State of Anarchy, have neither Law nor any other Rule except the Law of Nature . . . to go by."

The repudiation of established authority did not lead to anarchy. In rebellious communities throughout the thirteen colonies, people organized new institutions of government. Whatever the name — Committee of Correspondence, Committee of Safety, Committee of Inspection and Observation — some new source of local order was installed. Delegates from the local communities also came together in conventions at the county and provincial level. In the interim between the repudiation of British authority and the ratification of state constitutions (a period of five or six years in some cases), Patriot committees and conventions controlled the basic functions of government: law enforcement, conflict resolution, tax collection, and military procurement.

In some areas, the success of British military operations weakened or destroyed the power of the Patriot governments. The disintegration of armed resistance in New Jersey in 1776 allowed Loyalists to gain control of many communities. Many residents became increasingly resentful of the British troops and the Loyalists, but others accepted their control. When the British offered New Jersey civilians a free pardon, almost 3000 of them took oaths of loyalty to the Crown.

Some civilians attempted to avoid military conflict and remain neutral. Patriot officials in Connecticut jailed eighteen Farmington men for failing to defend nearby Danbury against a British raid in 1777. Initially suspected of being "inimical to America," these reluctant revolutionaries were interrogated and found to be "grossly ignorant of the present war with Great Britain." However, after some prodding, they "appeared to be penitent of their former conduct, professed themselves convinced since the Danbury alarm that there was no such thing as remaining neuters."

For many Americans, political allegiance was primarily a function of prudence rather than of passion. When the proximity of conflict made some choice unavoidable, they allied, or at least accommodated, themselves to whichever side held local control.

The Economic Crisis of War

Neither side could control the economy. The outbreak of war disrupted economic life at all levels, from the national to the local. Goods were scarce, prices were high, and no one was confident about the future. People struggled to adjust to a changing economic order, but were confronted with increasing disorder.

The main problem was obvious. Britain was no longer the favored trading partner that consumed or handled 40 percent of American exports, but had become the enemy that blockaded American ports and harassed American shipping. Some American traders continued to ship goods to other European countries and to the West Indies, but the British naval blockade made these ventures very risky. New England fishermen had no steady market for their catch; southern planters had fewer buyers for their tobacco, indigo, and rice. Merchants could not easily import finished goods. The decline in foreign trade also hurt those employed in the ship-building industry and other related trades — carpenters, ropemakers, coopers, and dock laborers. Especially in the larger port cities such as Boston, New York, and Philadelphia, the local economy almost collapsed. Hundreds of artisans and workers drifted off into the countryside to find work on farms or in small villages, joined the Continental Army, or, if they had Loyalist sympathies, departed with the British forces of occupation. The population of New York City declined from 21,000 in 1774 to less than half that number in 1783.

The war created a great internal demand for domestic goods. The interruption of importation gave renewed importance to the household manufacture of goods, especially of textiles. Women who had become accustomed to purchasing finished cloth from England found themselves once again spinning and weaving. The needs of the Continental Army swelled the demand for foodstuffs and other farm products. The war, however, inhibited the expansion of farm production even as it increased the demand. The army took away a significant proportion of farm laborers and livestock. The disruption of trade also made agricultural equipment difficult, if not altogether impossible, to replace. Throughout the war, demand almost always exceeded supply, and domestic production simply could not meet the needs of both the army and the civilian population.

The problem of expanding demand was compounded by an excessive paper money supply. The Continental Congress had little gold or silver with which to finance the war effort. Moreover, it had no power to impose taxes on the individual states, but could only make requisitions for revenue. In order to maintain political good will, the congress kept such requests moderate. Loans and gifts from foreign countries such as France and the Netherlands helped finance the war, but were not sufficient to keep up with government spending. The Continental Congress thus began to purchase supplies from farmers and artisans with paper currency, pledging to redeem the bills with future tax receipts. Since the congress issued far more paper money than it collected in coinage as taxes, however, people soon lost confidence in Continental currency. A growing fear that the bills would not be redeemed at face value led to a steady depreciation in their purchasing power. By 1777, it took 3 dollars in Continental bills to buy goods that cost 1 dollar in gold or silver. This ratio increased to 7 to 1 in 1778, 42 to 1 in 1779, 100 to 1 in 1780, and finally to 146 to 1 in 1781, when the currency ceased to have monetary value. The phrase "not worth a Continental" was a term of disparagement that was in common usage well into the nineteenth century.

The state governments contributed to the inflationary trend. To pay their own troops, they also issued large amounts of paper money. In Maryland, a bag of salt that cost $1 in 1777 was valued at $3900 three years later. The price of a bushel of wheat increased by a factor of 5000 in the same period. Most of this increase was "artificial"; it simply reflected the enormous increase in the supply of paper money. However, there was also a real shortage of goods. Even after an embargo had been placed on the export of foodstuffs from Philadelphia in

1779, the price of flour (in gold and silver) was still 113 percent above its prewar level.

This runaway inflation was disastrous for common people. Wage laborers in the cities suffered the most because wages lagged far behind the rising prices for goods. Few workers had cash reserves, and therefore most could not easily hold out for higher pay. They had to work at the going wage. Farmers were somewhat better off, but they did not enjoy the full benefit of rising prices for their foodstuffs. As one observer explained in 1778,

> . . . tho' the Produce of the Country Command more than six times its former Value, it is not Adequate to the Exorbitant prices you are Obliged to give for the Common Necessaries of life. . . . [By] the time an Article is consumed, it will have half a dozen Masters, each of whom has his profit which falls at last on the backs of the poor Farmer, or planter, who are the only people that bring produce to Market.

The debilitating effects of the spiral of currency depreciation and price inflation brought calls for control. At the end of 1776, delegates from the New England states met at Providence to set prices for domestic commodities and manufactured goods. A similar meeting at New Haven the following year established wage guidelines as well. Farmers, mechanics, and manufacturers were told to limit price increases to a maximum of 75 percent above the level of 1774, while tradesmen were allowed an increase of only 25 percent. These political controls were generally ineffective. Farmers refused to sell at the established prices. They either hoarded their goods or sold them in the black market to merchants with military contracts. To correct such abuses, the Massachusetts legislature passed an "Act to Prevent Monopoly and Oppression" in 1777. Three years later, however, the state government admitted that the law was a failure; it had led to "such a stagnation of Business and such a withholding of articles as had obliged the People to give up its measure or submit to starving."

The most effective price controls were often those imposed by the people themselves. Civilians did not commonly steal goods or ransack shops; they merely asserted their right to have necessary goods at what they considered a fair price. In 1779, a crowd of men and women in Philadelphia seized a merchant suspected of exporting flour that was desperately needed in the city. They made him sell the flour to city dwellers, and then they forced the town meeting to set wholesale and retail prices for thirty-two commodities. Crowds took similar actions in Boston and New York and in smaller towns. Popular enforcement of the "just price" was an old European tradition that had been followed many times before in the colonies. During the economic crisis of the Revolution, the just price had a special urgency and a widespread appeal.

The War Winds Down

Between 1779 and 1781, the American war effort was almost as feeble as the economy. After fighting to a military stalemate at the Battle of Monmouth

Courthouse in New Jersey in June, 1778, the British and American armies drew apart, and the war in the North came almost to a standstill. Even without dealing with the British, Washington had trouble controlling his army. In two consecutive winters, 1779–1780 and 1780–1781, Continental troops encamped at Morristown, New Jersey, mutinied. Cold, hungry, tired, and sick, these seasoned soldiers from New Jersey and Pennsylvania had had enough of army life. When they heard of the generous recruitment bounties being given to new enlistees, they declared their own enlistments were up. To restore morale, the Continental Congress granted concessions of back pay and new clothing. But, as a deterrent to future insurrections, Washington ordered several of the ringleaders shot.

More dramatic and disturbing was the notorious treachery of Benedict Arnold. Arnold had been one of the best and bravest young officers in the army — a daring leader in the capture of Fort Ticonderoga in 1775 and in the battle of Saratoga in 1777. His easily bruised ego suffered, however, when senior officers failed to give him proper credit for his combat exploits and when the congress passed him over for promotion. Like other gentlemen-officers, Arnold considered such slights an affront to his honor. He told the congress, "Honor is a sacrifice no man ought to make." His growing frustration with the army and with the congress (along with his straitened financial circumstances) led him to the ultimate dishonor. In September, 1780, he attempted to turn the American fortress at West Point over to the British. Arnold's plan failed, but the British still rewarded him with the generalship the Americans had denied him.

American officers were vitriolic in their abuse of Arnold, but their own loyalty to the cause was less than perfect. Many threatened to leave the army. The long-festering conflict over lifetime pensions surfaced again in 1779 and 1780. In a petition addressed to the Continental Congress, unhappy officers declared they would resign en masse if the congress did not meet their demands. Two years later, when the war was essentially over, a group of disgruntled officers hatched a plot at Newburgh, New York, to displace the civilian government altogether. Only an emotional entreaty by George Washington in March, 1783, stopped the Newburgh conspiracy from becoming a full-scale officers' revolt — and perhaps a military coup.

Meanwhile, the British shifted the focus of their war effort to the South. Intelligence reports from southern Loyalists led the British commanding officers to believe that the southern states were more politically and militarily divided, and thus more vulnerable, than the northern states. The presence of strong Indian tribes and restless slaves posed a nagging threat to the security of whites. In addition, lingering social and sectional antagonisms between lowland planters and backcountry farmers kept whites divided among themselves. The British hoped to exploit these divisions in order to gain control of one of the southern states and then use that state as a base of operations against the rest.

This strategy proved remarkably effective at first. After a successful amphibious assault on Savannah, Georgia, British regulars and Loyalist partisans quickly overcame rebel resistance and reestablished royal control over Georgia.

They found substantial support among the populace. Some 1400 Georgians swore allegiance to the king, and Loyalists organized twenty new militia units. The British then moved on to South Carolina. In December, 1779, a massive force of 8000 British troops under General Henry Clinton besieged Charleston, the most strategic port in the South. The American defenders under General Benjamin Lincoln resisted for a time, but a British naval bombardment forced them to give up. With the surrender of Charleston in May, 1780, the British captured more than 3400 Americans. Three months later, at Camden, South Carolina, British forces under General Lord Charles Cornwallis inflicted another stunning defeat on American troops. General Horatio Gates, the hero of Saratoga, led a hastily organized army of militiamen against the British, but the inexperienced Americans were no match for seasoned regulars. Panicking in the noise and confusion of battle, some militia units fled without firing a single volley. American casualties numbered over 1000. The American force essentially fell apart, and Gates was fortunate to escape with his life.

The war in the Carolinas quickly developed into a series of bloody and brutal battles between small groups of local Loyalist and rebel militiamen. For years, the Carolina backcountry had been an unstable, virtually ungoverned region. The outbreak of war had simply provided another cause — as well as a cover — for local conflict. Western yeomen who detested the eastern elite (who comprised the leadership of the Patriot movement) declared themselves loyal to the Crown. Other farmers who suffered at the hands of the Loyalists took up arms in nominal support of the revolutionary cause. Some people switched sides repeatedly. They fought less for principle than for plunder. One observer described South Carolina as "a piece of patch work, the inhabitants of every settlement, when united in sentiment, being in arms for the side they liked best, and making continual inroads into one another's settlements." The ferocity of civil conflict in the Carolinas reached a peak, both literally and figuratively, at the Battle of Kings Mountain in October, 1780. There, rebel militia units overwhelmed and destroyed a British force largely made up of local Loyalists. Under the guise of engaging in guerrilla warfare, neighbors settled old scores and righted old wrongs by committing new ones.

Nathanael Greene, a Rhode Island Quaker, emerged as the only successful American general in the South. Greene had previously been quartermaster-general for Washington's northern army. When he arrived in North Carolina to replace Gates in December, 1780, he expressed shock at the conditions of political and military chaos. "The word difficulty when applied to things here," he wrote to a fellow officer, "is almost without meaning, it falls so short of the real state of things." The American army in the South was "without discipline and so addicted to plundering that the utmost exertions of the officers cannot restrain the soldiers. Nor are the inhabitants a whit behind them." Greene immediately began to reorganize the army. He divided his forces into smaller, swift-moving units led by Daniel Morgan, Richard Henry "Light Horse Harry" Lee, Francis "The Swamp Fox" Marion, and himself. He realized that he could

THE WAR IN THE SOUTH, 1778-1781

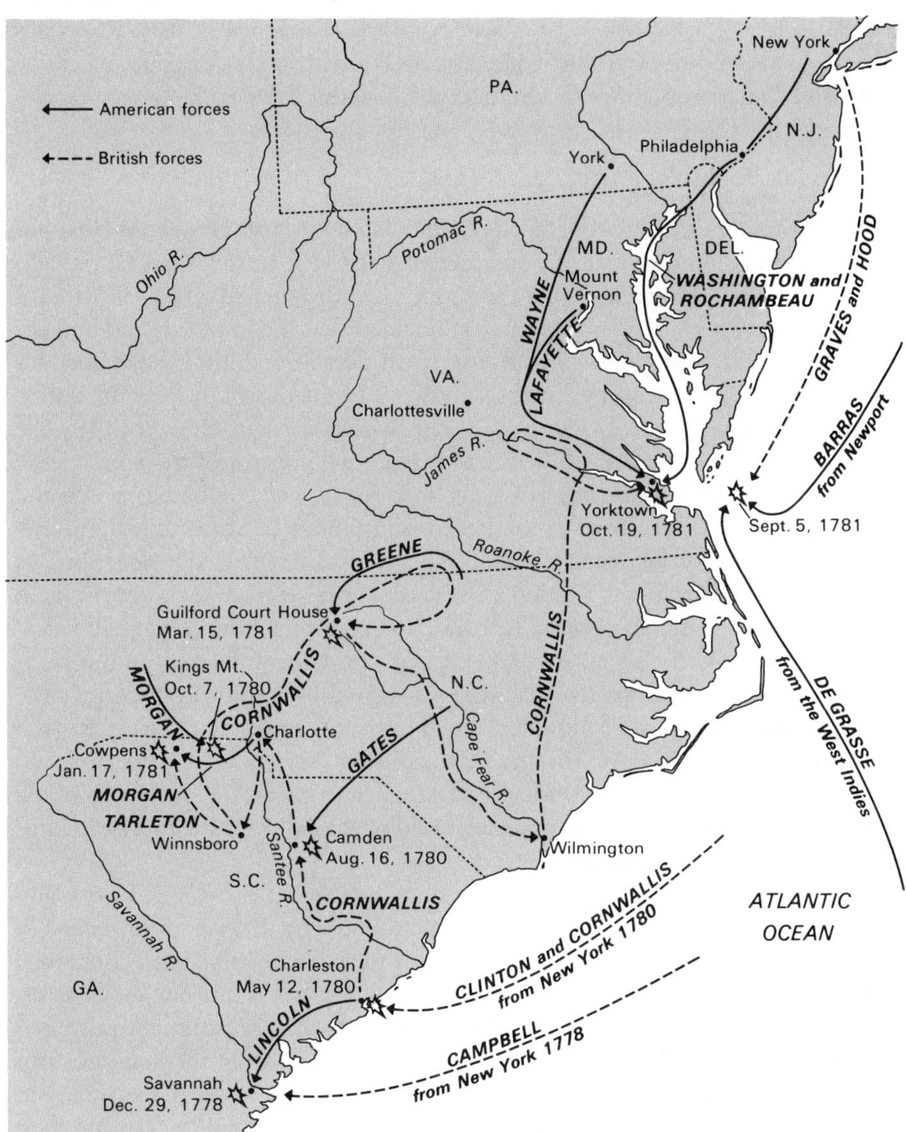

By early 1779, the British had captured the key ports of Savannah and Charleston, but could not control the southern backcountry, even though they received substantial support from Loyalist residents. Following eighteen months of guerrilla warfare and inconclusive military engagements, Lord Cornwallis led the main British army into Virginia. American and French forces surrounded these troops at Yorktown, forcing Cornwallis to surrender in October, 1781.

not defeat Cornwallis in a large-scale battle, so he adopted a strategy of harassment. The rebel forces led Cornwallis on an exhausting chase across the countryside. After a series of costly but nondecisive engagements, Cornwallis effectively gave up the Carolinas. He retreated northward to Virginia, where he established a garrison at Yorktown, near the mouth of the York River.

Victory at Yorktown and Peace Talks at Paris

Washington had for some time been planning an assault on New York City, but he changed his mind during the summer of 1781. He learned that Cornwallis had moved to Yorktown and that the French had dispatched a fleet of 30 ships and 3000 men to America. Washington decided that a combined land and sea operation could trap Cornwallis at Yorktown and force him to surrender his army. Accordingly, Washington took most of his army south, leaving only a small decoy force in the North to fool the British army in New York. By the end of September, the French fleet had established control of the Chesapeake Bay and was in place at the mouth of the York River. The combined French and American land force of over 16,000 men outnumbered Cornwallis's army by two to one. Cornwallis made a few feeble attempts to break the siege, but he soon realized fighting was fruitless. His troops were surrounded, cut off from reinforcements, vulnerable to artillery fire, and suffering from hunger and sickness. Cornwallis knew he had lost. On October 19, Cornwallis's second-in-command presented his sword to Washington's second-in-command, General Benjamin Lincoln. (Cornwallis himself claimed to be too "indisposed" to attend the surrender ceremonies.) At Yorktown, the British surrendered more than an army. They also lost their last hope of holding onto their American empire. As Lord North, the British prime minister, cried when he heard the news of Cornwallis's defeat, "Oh God, it is all over!"

Peace talks began in April, 1782, and on November 30, 1782, British and American commissioners signed a formal treaty. The British were weary of fighting the Americans and wary of French designs to reestablish a colonial empire in North America. As a result, they gave the Americans favorable terms. Both sides agreed to cease hostile action. The American government promised to curtail the confiscation of Loyalist property and to urge the state governments to return property already confiscated. The British kept Canada, but yielded all territory east of the Mississippi River and promised to evacuate their land and sea forces "with all convenient speed." They also recognized American fishing rights off the coasts of Nova Scotia and Newfoundland. Most important, and most obvious, they finally and formally recognized American independence.

The Americans achieved more than they had originally expected — or wanted. The thirteen American states assumed the status of a new nation in the eyes of the world. Victorious over England in war, they remained vulnerable to the diplomatic designs of the major powers. John Adams wrote:

> America has been a football between contending nations from the beginning, and it is easy to foresee, that France and England will both endeavour to involve us in their future wars. It is our interest and duty to be completely independent, and to have nothing to do with either of them, but in commerce.

Independence also had important domestic implications. It allowed Americans to establish their own government and determine their own course of development, free from wartime requirements and restraints. Peacetime would witness the real definition of independence, both for the individual states and for the American people as a whole.

8

The Parameters of Revolutionary Change

Americans did not fight for their independence merely to preserve their prerevolutionary society. They rejected the traditional social and political order inherent in monarchy and aristocracy. In its place, they sought to create a new society based on the republican principles of liberty and equality. By the end of the revolutionary era, the individual states and the nation were all established under republican constitutions that guaranteed free elections and representative government. Unlike most Europeans, American citizens could choose their leaders, from the local level to the national. That fact alone established the American Revolution as a remarkable and truly revolutionary political phenomenon.

The Revolution meant more than a change in the nature of political structures. The Declaration of Independence spoke broadly of human equality, of the right of all people to enjoy "life, liberty, and the pursuit of happiness." Those words held a powerful promise. A spirit of equality pervaded the independence movement, creating new expectations among people who had lived in positions of social and political inferiority. For many people the Revolution was a war fought on two fronts: they were struggling to win their freedom from Britain while striving for a new reality of equality within American society itself. The extent to which they achieved equality provides an important measure of the success and significance of the American Revolution.

A New Political Order

Even before the rebellious colonies officially declared their collective independence from Britain, they began drafting new constitutions. Having risen in arms against monarchical authority, the new states had to establish their own standards of republican political order. Most states completed the constitution-making process by 1778. Connecticut and Rhode Island simply made slight revisions to their colonial charters, which already provided for popular elections of state leaders, including governors and councillors. Pennsylvania adopted the most radical of the state constitutions. Its new government consisted of only one legislative body, with no governor or upper house. All legislative sessions were open to the public. Moreover, as a guard against the creation of an entrenched political elite, the Pennsylvania constitution prohibited legislators from holding office for more than four years out of any seven. Most other states were less democratic. Their constitutions embodied the principle of "balanced" government, with a governor and elite upper house as a check on any democratic excesses of the lower house.

Massachusetts experienced the most difficult transition from royal charter to state constitution. In 1778, in response to growing popular pressure, the assembly submitted a draft constitution to the town meetings for their consideration. People throughout Massachusetts raised strong objections to numerous sections of the proposed constitution, however, and a majority of towns refused to ratify it. The voters of the small western town of Greenwich rejected the constitution because, they explained,

> . . . Said Constitution and Form of Government . . . Entirely Divests the good People of this State of Many of the Privileges which God and Nature has Given them . . . and [gives] away that Power to a few Individuals, which ought forever to Remain with the People inviolate, who Style themselves free and Independent.

Two years later, another constitutional convention sent out a revised draft to the towns, and this time the constitution won popular approval — although by a slim margin.

With the ratification of the Massachusetts constitution in 1780, all the states had defined their principles of government. Taken together, these state constitutions embody the range of new political possibilities envisioned by the American voters and their elected leaders. They also reveal how some of the old political restrictions still endured.

The new state constitutions allowed for greater popular participation in the political process. On one level, this movement toward increased voter representation was inspired by the republican emphasis on political equality and the virtue of the people. More important, this change was dictated by two crucial political realities of the Revolution. First, elite (and largely eastern) leaders of the Patriot cause needed the support of the common people in the rural interior and western backcountry. Second, common people took to heart the democratizing sentiments of the independence movement. Urban mechanics and hinterland farmers had been extremely active in revolutionary committees and conventions. They had asserted themselves politically as they had never before done (or been allowed to do). As a writer calling himself "Democritus" told the voters of central Massachusetts, "If you would be well represented, choose a man in middling circumstances as to worldly estate, if he has got it by his industry so much the better, he knows the wants of the poor, and can judge pretty well what the community can bear of public burdens." More people — from more levels of society — began taking part in political decision-making.

This new political atmosphere contrasts strikingly with that of the prerevolutionary era. In colonial New Hampshire, well-to-do men from coastal communities controlled most of the thirty-four seats in the assembly in 1765. By 1786, the house had swelled to eighty-eight members. Over half of the new representatives came from inland farming communities, and an equal proportion were yeomen. A similar pattern prevailed in South Carolina. In 1772, only forty-eight men sat in the colonial assembly, almost all of them lowland planters. By 1775, the revolutionary provincial congress had four times as many members, and a third of them came from the western upcountry. Members of the eastern elite exerted enough control over the drafting of the state's constitution in 1778 to guarantee their region disproportionate representation, but upcountry representatives held 40 percent of the seats in the house in 1785. A Virginian noted with approval the change in his state. The new House of Burgesses, he observed, was "composed of men not quite so well dressed, nor so politely educated, nor so highly born as some Assemblies I have formerly seen. . . . [But] They are the people's men (and the People in general are right)."

Many framers of state constitutions did not place such faith in the people and were unwilling to throw government open to the masses. Several states imposed substantial property requirements on candidates for higher elective offices. In Massachusetts, the governor had to own property worth at least £1000; in Maryland, the requirement was £5000, and in South Carolina, £10,000. Property requirements for state senators and assemblymen were lower, but were still restrictive enough in some states to discourage common people from seeking higher office. In Maryland, only 11 percent of the state's white males qualified to hold seats in the lower house, and just over 7 percent were eligible for the upper house.

Property requirements were also applied to the electorate in most states. Only Pennsylvania and Georgia allowed all taxpayers to vote. All the other states retained some property qualification on the right to vote. In North Carolina, all taxpayers could vote for representatives for the lower house, but only men who owned at least 50 acres of land could vote for senators. The Massachusetts property requirement was actually slightly higher after the Revolution than before. Elsewhere, the property requirements were generally lower than they had been before the Revolution. The fact that any such requirements were retained gave constitutional sanction to the traditional assumption that only those with a financial stake in the society could be safely trusted with its management.

The Status of Women

The state constitutions upheld another traditional assumption — that only men could vote and hold public office. This restriction was partly a natural extension of the property requirement. Because most married women lacked the legal status to own property, they did not have the required stake in society. Also, the exclusion of women from political life had deep roots in American society. It was part of a complex web of laws, habits, and traditions, all of which reinforced the pervasive opinion that women were fit only to function in the private sphere of the family. As Timothy Dwight noted with approval, "Women in New England are employed only in and about the house, and in the proper business of the sex."

Abigail Adams was one New England woman who refused to accept this narrow definition of female identity. This remarkably astute and articulate woman spoke for thousands of others in suggesting a higher status for women in the new republican society. In 1782, she wrote her husband, John Adams:

> Patriotism in the female sex is the most disinterested of all virtues. . . . Even in the freest countries our property is subject to the control and disposal of our partners, to whom the Laws have given a sovereign Authority. Deprived of a voice in Legislation, obliged to submit to those Laws which are imposed upon us, is it not sufficient to make us indifferent to the public Welfare? Yet all History and every age exhibit Instances of patriotic virtue in the female sex; which considering our situation equals the most Heroic.

During the Revolutionary War, Abigail Adams and other American women performed heroic services, both in the military and at home. Hundreds of women traveled with the rebel army, mostly as "camp followers." They provided cooking and medical care for the troops and endured the same hardships as the men. Other women served as spies and messengers, and a few actually took part in combat. Deborah Gannet Sampson of Middleborough, Massachusetts, disguised herself as a male soldier and served in the army for over a year until she was wounded. On the home front, ordinary women carried on a variety of vital tasks. With husbands and sons away at war, some wives had to assume control of farms and businesses. While John Adams was serving in the Continental Congress, Abigail Adams wrote proudly of herself as a "farmeress."

Abigail Adams made clear her political sensitivities. In a well-known and remarkably revealing series of letters to her husband during the spring of 1776, she urged the men in the Continental Congress to "Remember the Ladies, and be more generous and favorable to them than your ancestors." She was not asking for full political equality for women or even for the right to vote; she merely asked that the lawmakers in the new republic provide women with better legal protection, especially against the power of their husbands. Taking up the language of the Revolution, she added:

Abigail Smith Adams and John Adams. These portraits, done by Benjamin Blyth in 1766, show this remarkable young couple in the early years of their marriage. The artist has captured the self-confidence and strength of character of Abigail, who would challenge her husband ten years later to "remember the ladies" when constructing the legal and political systems of the new republic. (*Source: Massachusetts Historical Society.*)

> Remember all Men would be tyrants if they could. If particular care and attention is not paid to the Ladies, we are determined to foment a Rebellion, and will not hold ourselves bound by any Laws in which we have no voice, or Representation.

Her husband responded with jocular condescension:

> As to your extraordinary code of laws, I cannot but laugh. Depend upon it, We know better than to repeal our Masculine systems. Although they are in full Force, you know they are little more than theory . . . and in Practice you know We are the subjects. We have only the name of Masters.

A month later, in a letter to another man, John Adams wrote seriously about the perils he saw in revolutionary political change:

> Depend upon it, Sir, it is dangerous to open so fruitful a source of controversy and altercation as would be opened by attempting to alter the qualifications of voters; there will be no end of it. New claims will arise; women will demand a vote; lads from twelve to twenty-one will think their rights not enough attended to; and every man who has not a farthing, will demand an equal voice

> with any other, in all the acts of state. It tends to confound and destroy all distinctions, and prostrate all ranks to one common level.

Like other conservative leaders of the Patriot movement, John Adams feared the revolutionary spirit he had helped create. If all people took the radicals' rhetoric to heart, he saw enormous implications for political and social change.

Events in New Jersey clearly revealed the potential for such change. The state's 1776 constitution defined voters as "all free inhabitants" who met the residence and property requirements. The language of the constitution was vague enough to allow widows and spinsters with £50 of property to claim the right to vote. Women who met the requirements took part in local elections from the 1780s through the early years of the nineteenth century, and some New Jersey men came to take their participation for granted. Other men objected strenuously and launched a campaign to eliminate women's suffrage. Calling himself a "Friend to the Ladies," one traditionalist appealed to their emotional ties to men. "How will an obedient daughter dare to vote against the sentiments of her father," he asked "and how can a fair one refuse her lover?" Another man put the matter more bluntly: "Women generally, are neither by nature, nor habit, nor education, nor by their necessary condition in society fitted to perform this duty with credit to themselves or advantage to the public." The conservative forces eventually triumphed. In 1807, the New Jersey legislature passed a statute restricting suffrage to white males.

Despite continuing legal restrictions, the era of the Revolution did bring about substantial positive change in the situation of women. Like Abigail Adams, many women grasped the egalitarian implications inherent in revolutionary ideology. In 1790, Judith Sargent Murray of Gloucester, Massachusetts, published an iconoclastic essay, "On the Equality of the Sexes" (which was originally written in 1779). Comparing women to men in terms of mental ability, she found women equal or superior to men in imagination and memory and inferior in reason and judgment — but only because females lacked formal education and intellectual training. In other writings, she stressed the importance of equality and mutuality in marriage: "Mutual esteem, mutual friendship, mutual confidence, *begirt about by mutual forbearance*." Another woman writer, calling herself "Matrimonial Republican," echoed Murray's sentiment: "I object to the word 'obey' in the marriage-service. . . . The obedience between man and wife, I conceive, is, or ought to be mutual."

These republican sentiments reflected subtle changes that had occurred in the reality of family life. Especially in the northern states, women who reached maturity after 1770 were less likely to marry and have large families than were their prerevolutionary counterparts. This phenomenon was the result of several demographic, economic, and cultural factors. In some older communities, the increasing scarcity and cost of farmland caused many young men to migrate westward. As a result, a surplus of single women soon existed in eastern towns. In Hingham, Massachusetts, 15 percent of the daughters born to one group of parents between 1721 and 1760 did not marry. Also, many women married at an older age and had fewer years in which to bear children. In Pennsylvania, Quaker women born after 1755 had an average of five children in their lifetimes, two fewer than Quaker women born before 1730; moreover, many of these younger Quaker women stopped having children before they reached menopause, which was indicative of a conscious decision to limit family size. Similar changes took place in several New England communities in the last quarter of the eighteenth century.

The decline in family size that began around the time of the Revolution was the first step in a long-term trend that lasted until the twentieth century. It coincided with a shift from patriarchal notions of marriage relationships to more egalitarian ones. Husbands were less likely to insist on having more children, and wives were more likely to refuse.

A new emphasis on "republican motherhood" gave women both greater responsibility and greater credit for raising virtuous citizens of the American republic. A list of "Maxims for Republics" pointed out: "Some of the first patriots of ancient times, were formed by their mothers." And John Adams wrote his daughter: "You, my dear daughter, will be responsible for a great share of the duty and opportunity of educating a rising family, from whom much will be expected."

Motivated by republican ideology, social responsibility, and personal ambition, many upper-class women sought greater public responsibilities. In 1797, Isabella Graham founded the New York Society for the Relief of Widows with Small Children. Her daughter, Joanna Graham Bethune, organized an Orphan Asylum Society. During the 1820s, mothers in hundreds of American towns and villages formed maternal associations to encourage Christian childrearing. Similarly, moral concerns led women in New York City to found the Female Moral Reform Society in 1834. Within six years, the nationwide American Female Moral Reform Society had over 500 chapters. Its members set up boarding houses for prostitutes, runaway girls, and working women. The reformers' religious activism ultimately propelled them into the political arena as they pressured city authorities to close down brothels.

The enhancement of the female role in the family and society led to calls for better education for women. The enrollment of girls in public schools increased dramatically in the postrevolutionary era. Private female academies supplemented the educational opportunities for young women from middle- and upper-class families. Although women still could not obtain academic training at colleges like Harvard, Princeton, and Yale, many received the same basic education available to the vast majority of men. By the 1820s, many graduates of female academies had become teachers. Poorly paid in relation to their male counterparts, they nonetheless enjoyed a substantial degree of economic and social independence. Through education, women had broken another of the invisible social chains that had traditionally bound them to the sphere of home and family.

Throughout the first half of the nineteenth century, however, the dominant cultural ideology still confined most women to the private sphere of the household. In the political sphere, only men were created equal. That observation was made — and its premises challenged — by American feminists at a meeting in Seneca Falls, New York, in 1848. Significantly, when those feminists drafted their Declaration of Rights and Sentiments, they self-consciously adopted the language of the Declaration of Independence.

The Contradiction of Slavery

All men were not equal, however. The continued enslavement of blacks ran counter to the egalitarian sentiments expressed in the Declaration of Independence and other revolutionary writings. Even free blacks had a status that was ambiguous at best and, in some cases, almost as degrading as slavery itself. Many of the whites who led the fight for freedom from Britain were fearful of, or hostile to, the implications of freedom and equality for blacks. Throughout the revolutionary era, the question of slavery and racial inequality was one of the most perplexing problems, practically and ideologically, that faced the independence movement.

The issue of military service for blacks provided an early indication of white racial attitudes. A few blacks served in the Massachusetts militia units at the battles at Concord and Bunker Hill, and during the summer of 1775, they remained with the American force besieging Boston. But Edward Rutledge, a delegate to the Continental Congress from South Carolina (where blacks constituted a majority of the population), demanded that blacks be excluded from military service. Once blacks were armed to fight the British, he argued, they would turn instead on their white masters. In November, 1775, he convinced both Washington and his colleagues in the congress to keep all blacks, free or slave, out of the Continental Army.

Ironically, the British had already taken action that would force the Americans to change their policy toward blacks. Lord Dunmore's successful appeal to Virginia slaves prompted Patriot leaders to counter the British offer. By 1778, all the states except Georgia and South Carolina allowed at least some blacks to serve in the military, either as enlistees or as substitutes. Massachusetts and Rhode Island formed all-black regiments. Black soldiers and sailors eventually accounted for approximately 5000 of the 300,000 Americans who took up arms for the Patriot cause. Some blacks who entered the army as slaves gained their freedom as a reward.

During the Revolution, freedom for all blacks became an issue of public debate. Jefferson pointed to the evils of slavery in his original draft of the Declaration of Independence. In his long list of grievances against the king, the Virginia slaveowner argued that George III had

> . . . waged cruel war against human nature itself, violating its most sacred rights of life and liberty in the persons of a distant people who never offended

American Soldiers, 1782. This drawing by a French army officer, Jean Baptiste de Verger, includes a black infantryman from a Massachusetts regiment. Blacks served in the military forces of many northern states and some southern ones as well, often receiving freedom as their reward for loyalty to the Patriot cause. (*Source: The Granger Collection.*)

> him, captivating & carrying them into slavery in another hemisphere. . . . Determined to keep open a market where *Men* should be bought & sold, he has prostituted his negative for suppressing every legislative attempt to prohibit or restrain this execrable commerce.

With the outbreak of war, Jefferson continued, the king (through Lord Dunmore) encouraged slaves to flee from their masters and join the British forces in order to "purchase that liberty of which he has deprived them, by murdering the people on whom he has obtruded them."

This indictment of the king was historically flawed. For a century and a half, many American colonists — including Jefferson himself and his fellow planters in Virginia — had voluntarily increased their reliance on slave labor; they could hardly argue that slavery had been "obtruded" on them. Jefferson's argument was also politically awkward. Few slaveowners admitted that black people enjoyed "sacred rights of life and liberty." At a time when the unity of the independence movement was so fragile and so crucial, Jefferson's fellow delegates in the Continental Congress were not willing to antagonize the southern states, especially Georgia and South Carolina. They deleted the reference to slavery from the final draft of the Declaration and ignored the issue of slaves' rights for the duration of the war.

Abolition and Manumission

At the state level, lawmakers confronted the contradiction of slavery more directly. During the Revolution, all the states except Georgia and South Carolina made provisions for freeing slaves. In Massachusetts, the state supreme court ruled slavery unconstitutional in 1783. In making his charge to the jury in the landmark *Quock Walker* case, Chief Justice Thomas Cushing acknowledged that the state constitution did not explicitly outlaw slavery. But Cushing argued that slavery was a "usage which took its origin from the practice of some of the European nations and the regulations for the benefit of trade of the British government." In a new revolutionary society, he concluded, such old usages no longer had a place:

> Sentiments more favorable to the natural rights of mankind, and to that innate desire for liberty which heaven, without regard to complexion or shape, has planted in the human breast — have prevailed since the glorious struggle for our rights began. And these sentiments led the framers of our constitution of government . . . to declare — *that all men are born free and equal;* and that *every subject is entitled to liberty*. . . . The court are therefore fully of the opinion that perpetual servitude can no longer be tolerated in our government.

Many other northerners had major reservations about the immediate abolition of slavery. James Winthrop of Massachusetts (a descendant of the Puritan founder) declared that, because of the court's decision in the *Quock Walker* case, "a number of citizens have been deprived of property formerly acquired under the protection of law." Winthrop had gone directly to the heart of the problem — the conflict between property rights and human rights that the movement for abolition had stirred up. Lawmakers in the northern states of Pennsylvania, New York, Connecticut, and Rhode Island avoided directly confronting this problem by providing for gradual emancipation. The Pennsylvania statute of 1780, for instance, opened with a preamble (written by Thomas Paine, the radical pamphleteer) that spoke of "removing, as much as possible, the sorrows of those, who have lived in undeserved bondage." But the law did not free slaves already in bondage; it only guaranteed freedom for slaves born after 1780 — and then only after they had served their mothers' masters for twenty-eight years.

Elsewhere, in New Jersey, Delaware, Maryland, Virginia, and North Carolina, state legislatures passed laws facilitating manumission, the voluntary freeing of slaves by their masters. Such laws did not force slaveowners to give up their property; they simply allowed (and perhaps encouraged) them to do so. Hundreds of masters responded to the opportunity. In Virginia alone, the number of free blacks rose from 2000 in 1782 (the year the state's manumission legislation was passed) to 12,000 in 1790, 20,000 in 1800, and 30,000 in 1820. As Table 8.1 shows, there were more than 77,000 free blacks in the Chesapeake region in 1810, and they comprised a larger portion of the total population there than in any other region in the country.

The movement for abolition and manumission reflected economic circumstances as well as ideological commitments. In the northern states, slaves were concentrated in the cities, where they worked primarily as household servants

TABLE 8.1
Black Population, by Region, 1790–1810

	1790			1800			1810		
	Slaves	Free Blacks	Free Blacks as % of Total Population	Slaves	Free Blacks	Free Blacks as % of Total Population	Slaves	Free Blacks	Free Blacks as % of Total Population
New England	3,886	13,156	1.3	1,340	17,317	1.4	418	19,488	1.3
Mid-Atlantic	39,284	11,153	1.1	34,471	29,337	2.1	26,663	55,668	2.8
Chesapeake	405,350	24,708	2.3	457,584	47,979	3.7	508,197	77,633	5.4
Lower South*	252,177	7,649	1.0	396,267	12,479	0.9	647,191	27,746	1.4

SOURCE: *Population of the United States in 1860* (U.S. Census Bureau, 1864), 600–601.
*Figures for Lower South include the addition of Mississippi (1800) and Louisiana (1810).

By 1810, free blacks constituted 13 percent of the total black population. Most free blacks lived in the mid-Atlantic and Chesapeake regions; only a small proportion of the total black population in the lower South was free. Planters in that region imported more than 250,000 African slaves between 1775 and 1807, when congressional legislation ended American participation in the Atlantic slave trade.

for wealthy families. Before 1750, slaves had also worked in artisans' shops, but the use of slave labor in the skilled trades had gradually declined. Especially during the period of economic uncertainty preceding the American Revolution, master artisans and small manufacturers preferred to hire free laborers who could be readily dismissed. The decreasing significance of slave labor made it relatively easy for northern legislators to rectify a moral wrong without creating economic disruption.

In the Chesapeake region, economic factors similarly lay behind seemingly progressive policies. Beginning in the 1760s, many planters shifted production from tobacco to grain. Wheat and corn not only brought a higher price than tobacco; their cultivation also required less labor. Since planters therefore needed smaller labor forces, it made economic sense to free some slaves. Other masters, who were reluctant to let their slaves go for nothing, allowed them to hire themselves out for wages during slack times in the season. Many slaves then purchased their own freedom with the income from their work.

In 1800, a group of slaves in Virginia attempted to gain their freedom in a more direct fashion. Led by blacksmith Gabriel Prosser, this group planned a surprise attack on Richmond, the state capitol. Their objectives were to capture the city, to seize the store of arms there, and to take the governor prisoner. They assumed other slaves would then rise up in a widespread insurrection. Prosser and his followers vowed to "Wade to our Knees in blood sooner than fail in the attempt." As one of the conspirators later confessed, he was prepared to "kill White people stoutly . . . I will fight for my freedom as long as I have breath, and that is as much as any man can do."

The leaders of this group already enjoyed a relative degree of freedom. Like Prosser, they were artisans who worked independently and traveled extensively in the Virginia countryside. They took advantage of their mobility to talk with plantation slaves at Sunday barbecues and religious meetings. The Christian religion played an important part in slaves' lives as a source of inspiration and liberation. White masters and ministers preached submission ("Slaves obey your masters"), but slaves found a different lesson in the Bible, which was the power of redemption. One of Prosser's allies drew upon this message to inspire recruits. He said he "had heard in the days of old, when the Israelites were in Servitude to King Pharoah, they were taken from him by the power of God — & carried away by Moses."

The insurrectionists were also inspired by the ideology and rhetoric of the revolutionary era. One slave declared that he was committed "to fight for his Country," just as white Virginians had fought for theirs:

> I have nothing more to offer than what General Washington would have had to offer, had he been taken by the British officers and put on trial by them. I have ventured my life in endeavoring to obtain the liberty of my countrymen, and am a willing sacrifice to their cause.

St. George Tucker, a white lawyer, acknowledged the slaves' political awareness. The slaves who had fled to join Governor Dunmore and the British in 1775 "fought [for] freedom merely as a good," he noted, "now they also claim it as a right." Although it was unsuccessful, Gabriel's insurrection made clear the pervasiveness and power of revolutionary ideas among slaves. Unfortunately for blacks, it also caused whites to conclude that they needed to keep tighter control over the potential rebels in their midst.

Rice planters in South Carolina and Georgia had stood firmly against emancipation and manumission. Following the War of Independence, they imported more than 250,000 new slaves from Africa. The emergence of the cotton economy dealt the deathblow to the movement to end slavery. Eli Whitney's cotton gin, invented in 1793, provided the crucial technological boost for an agricultural boom. Unlike the long-staple cotton grown only on the Sea Islands of Georgia and South Carolina, which could be cleaned easily enough, the short-staple, or upland, cotton grown on the mainland was hard to clean. Separating the seeds from the fiber was done by hand and was a slow and laborious task. Whitney's cotton gin made it possible to clean cotton bolls almost as fast as they were picked, and cotton production soared. In 1792, the South produced 6000 bales of cotton. By 1800, production had multiplied by twelvefold, to 73,000 bales; it soon doubled again, to 146,000 bales in 1805 — for a total increase of over 2300 percent in less than a quarter-century. In 1806, raw cotton sold for 18 cents per pound. By planting four acres of cotton on newly cleared land, yeomen farm families could make a net profit of $200 per year. Substantial profits from cotton farming financed the settlement of new territories. In 1790,

Advertisement for Slaves (Charleston, South Carolina, 1787). The United States Constitution of 1787 declared that there could be no congressional legislation to outlaw the international slave trade until 1807. In the meantime, slaveowners from the lower South, who were responsible for the inclusion of this provision in the Constitution, imported hundreds of thousands of new slaves from Africa and the West Indies. (*Source: American Antiquarian Society.*)

100,000 settlers lived in the future states of Kentucky (1792) and Tennessee (1796). By 1820, the number of residents in these states had soared to 987,000.

Just as the tobacco boom encouraged racial slavery, the cotton boom perpetuated and extended it. The spectacular growth of the cotton economy gave new importance to the slave system by increasing the demand for field laborers. The slave trade from Africa officially ended in 1808, but an "internal" slave migration disrupted the lives of millions of Afro-Americans during the early nineteenth century. Thousands of white planters forced their slaves to accompany them to the upcountry regions of South Carolina and Georgia or to the new states of Mississippi (1817) and Alabama (1819). Some blacks were moved to the "cotton belt" in family units, as the chattels of migrating sons and daughters of established Chesapeake and Carolina planters. Many others, especially strong young men and boys, were sold to migrating neighbors or to slave-traders. Torn away from family and community, these blacks were condemned to the hard labor of cutting new plantations out of forested wilderness. For most Afro-Americans, the rise of the cotton economy was an unmitigated disaster.

The Limits of Black Freedom

Many free blacks faced discouraging prospects in the postrevolutionary era. Former slaves usually had to move away from the plantation regions, even though family ties made them reluctant to do so. Plantation whites regarded free blacks with hostility and anxiety; they believed the presence of former slaves would encourage widespread restlessness or even rebellion among enslaved blacks. Some southern states forced free blacks to leave. Others required them to register with state authorities, or to have white guardians. Free blacks had to carry certificates of freedom; any black person at large without such a document was presumed to be a slave and could be returned to bondage. Some free blacks moved west, especially to the newly opened Northwest Territory, where slavery was outlawed by Congress in 1787. Most free blacks migrated to cities in the east, where they could make a living without having to buy land.

Wherever they went, free blacks found that personal liberty did not guarantee social or political equality. A few notable blacks, such as the poet Phillis Wheatley and the shipowner Paul Cuffee, achieved prominence and even a degree of prosperity. Most free blacks remained propertyless and poor. Some actually suffered a decline in occupational status. Legal restrictions in the southern states prohibited blacks from engaging in certain lines of work, ranging from peddling to typesetting. Those who had been artisans on plantations found themselves shut out of their skilled trades by whites who objected to working with, or feared losing their jobs to, blacks. Employers often played free blacks off against white workers, thus intensifying tensions among the urban working classes.

Free blacks suffered other forms of discrimination as well. Many white churches in the North excluded blacks, forcing them to establish their own congregations. Although some religious groups and abolitionist societies established schools for blacks, most cities and towns refused to provide public funds for black education. Blacks were also denied participation in the political process. Most state constitutions did not initially exclude blacks from voting, but lawmakers gradually added restrictions that curtailed black suffrage. The 1807 New Jersey law that disfranchised women also specifically denied blacks the right to vote. In the early part of the nineteenth century, blacks were also disfranchised by statute in Washington, D.C. (1802), Maryland (1810), Tennessee (1834), and North Carolina (1835). In other states, property or residence requirements effectively precluded black suffrage.

Some free blacks turned exclusion into strength. Led by Richard Allen, a talented Methodist clergyman, a group of northern blacks established the African Methodist Episcopal Church. By 1820, the black members of this denomination numbered 4000 in Philadelphia and 2000 in Baltimore. AME and other black churches became an important institutional base for the development of black identity and leadership in the nineteenth century; in fact, they were critical to the rise of the abolitionist movement.

The Crisis of the Postwar Economy

All Americans felt the economic consequences of the Revolution long after the military conflict ended. Many merchants and agents reaped windfall profits from military supply contracts. "Fellows who would have cleaned my shoes five years ago, have amassed fortunes, and are riding in chariots," complained Boston's James Warren in 1779. Such envious sentiments were echoed by many others during the 1780s. Most Americans had contended with high prices, scarce goods, and depreciated currency during the war years. The war's end left them economically strained and still vulnerable. The first decade of peace was a time of economic readjustment, which led to increased social instability and eventually to open conflict.

For a brief period after the war, the economic prospects seemed reasonably bright to Americans. Leading London merchants had convinced the British government to permit duty-free imports and exports between Britain and the former colonies. Eager to replenish depleted inventories and make quick profits, American merchants immediately resumed trading with their British counterparts. Other European countries had never fully replaced Britain as customers for American exports. In addition, no one offered easier credit than British mercantile houses.

The renewed trade relations with Britain differed significantly from the prewar situation. The seemingly generous free-trade policies enacted by Parliament applied only for the British Isles themselves. The British colonies in the West Indies, which had formerly been such lucrative markets for mainland exports, were declared off-limits to American vessels; all products exported to the island colonies had to be carried in British ships. This loss of the West Indian market dealt a severe blow to the American economy. John Adams noted in 1783 that "the commerce of the West India Islands is part of the American system of commerce. They can neither do without us, nor we without them." Two years later, a New England merchant complained of the effects of the British trade regulations: "Our West Indies trade is much against us, being shut off from the English Islands and closely restricted amongst the French." By controlling American trade with the islands, Parliament reduced the new nation to a state of economic dependency. Having lost the war, the British were winning the peace.

As Table 8.2 indicates, American commerce suffered significantly during the immediate postwar era. The value of exports to Great Britain and the British West Indies had averaged $7.5 million per year between 1771 and 1773; this figure dropped to an average of only $5.8 million per year between 1785 and 1787. Even though the proportion of American exports to nations other than Great Britain rose from 30 percent to 50 percent between these two periods, the increase was not enough to maintain the prewar per capita level of exports or a favorable balance of trade. As one merchant observed ominously in 1783, "The importation of European goods has been so considerable, within the last six months, that the exportable produce of the states probably will not equal it in less than three years." Other American merchants realized the danger of the trade imbalance, but in many cases it was already too late. When British creditors called for payment in silver and gold, overextended American debtors soon exhausted their supply of specie. Some merchants defaulted on their debts; others had to scramble to stay solvent.

State governments also had to settle with their creditors. During the war, the states had borrowed huge sums of money and had printed vast amounts of paper currency. The coming of peace brought with it calls for repayment — in hard currency. In order to pay off the debt, state governments increased the tax burden of their citizens. Most states imposed higher import duties and land taxes, and some required that these taxes be paid in hard currency. Public indebtedness thus compounded the growing economic pressure in the private sector.

From Economic Oppression to Armed Insurrection

The crisis of the postwar economy reached its most disturbing climax in New England, especially in Massachusetts. In 1784, the first full year after the war, New England merchants purchased British imports valued at almost £600,000 — about the same amount they had imported annually in the decade before the Revolution. But the postwar drain of specie to British creditors cut sharply into the New England economy. Massachusetts, the leading mercantile state in the region, had only £150,000 in specie still in circulation by mid-1784; several times that amount had already gone back to Britain, and even more was still owed. Despite the specie shortage, the state government required the pay-

TABLE 8.2
American Exports, 1770–1793
(in millions)

Period	To Great Britain	To British West Indies	To Others	Total	American Population
1771–1773	$5.3	$2.2	$3.2	$10.7	2.4
1785–1787	4.4	1.4	5.8	11.6	3.6
1792–1793	4.3	1.2	5.5	11.0	4.2

SOURCE: Compiled from Gordon C. Bjork, "The Weaning of the American Economy: Independence, Market Changes, and Economic Development," *JEH* 24 (1964), Tables 2 and 3.

British restrictions on American commerce, especially on trade with the West Indies, cut exports, despite the efforts of American merchants to find markets in other countries. The value of American exports stayed at $11 million between 1770 and 1793, while the population increased by 75 percent. Per capita imports and living standards fell, contributing to social and political discontent.

ment of taxes in hard currency. Merchants also began asking their customers for payment in specie. Facing pressure from their British creditors, urban merchants squeezed hard currency out of rural traders, who in turn looked to farmers as a source of specie.

The farmers at the end of this chain of debt were unable to pay in hard currency, and were unaccustomed to doing so. Rural people generally used paper currency or farm products — grain, wool, cloth, and other homemade goods — to cover their debts. Moreover, the culture of the countryside traditionally favored flexible credit relationships, and rural creditors often carried unpaid debts over a period of two or three years.

The postwar economic pressure imposed by local merchants created anxiety and anger among poor farmers. They dreaded the prospect of debt proceedings that could cost them their land, and therefore their livelihood. Farmers in the small western Massachusetts town of Conway wrote that the

> . . . *mortgage of our farms,* — we cannot think of, with any degree of complacency. To be *tenants* to *landlords,* we know not who, and pay rent for lands, *purchased with our money,* and converted from howling *wilderness,* into fruitful fields, by the *sweat of our brow,* seems . . . truly shocking.

The Conway farmers' petition addressed an urgent problem. Legal suits over debt skyrocketed in their own Hampshire County. Between August, 1784, and August, 1786, the county court heard 2977 debt cases, an increase of 262 percent over the 1770–1772 period. These debt proceedings involved almost a third of the adult male citizens in that county; similar proportions of residents were involved in other rural regions of New England. Prosecution for debt did not always lead to confiscation of an entire farm, for most debt cases involved fairly small sums. Defending debt cases drained farmers of time and money, however, and the fear of confiscation always loomed in the background. As the governor of Vermont noted in 1786, "law suits [have] become so numerous that there is hardly money sufficient to pay for entering the actions, not to mention the debts."

In Massachusetts, frustrated farmers turned to their government for relief. They sought legislation to allow them to pay their taxes and debts with paper currency rather than specie. Representatives of the depressed western counties introduced bills for a new issue of state currency and for the regulation of legal fees. Although the legislation passed in the lower house, it was defeated in the senate, where eastern creditors held disproportionate influence.

Rebuffed by their government, the people of the poorer western towns resorted to the same measures they had used against the British in 1774. They met in extralegal conventions to condemn the government, and then they forcibly closed the county courts. As at the time of the Revolution twelve years earlier, the farmers did not initially seek the overthrow of the government; they only wanted a redress of their grievances and a restoration of their just rights. When they spoke of the "Suppressing of tyrannical government," eastern politicians clearly understood the historical allusion. As archconservative Fisher Ames commented, "The people have turned against their teachers the doctrines, which were inculcated in the late revolution."

Under the leadership of Captain Daniel Shays, a respected revolutionary army officer, the Massachusetts rebels rose in armed insurrection. With a force that on occasion numbered as high as 2000 men, they constituted a formidable threat to the authority of the state government. State officials could not count on the loyalty of the militia; many militiamen refused to fight against their neighbors, especially in the western counties where rebel sentiment was intense. Nervous eastern merchants contributed money for recruitment bonuses and supplies, however, permitting the state government to raise an effective fighting force. Governor James Bowdoin and his council also made a secret request to Congress for military assistance. Congress agreed to supply 1340 men at a cost of more than $500,000.

By the time the national force was organized, Shays's Rebellion was over. Poor leadership, inadequate supplies, and freezing weather reduced the effectiveness of the rebel force in the winter of 1786–1787. By the spring, the state troops had broken up the rebel army into small bands. Most of the Shaysites surrendered and signed oaths of loyalty to the state. In the following year, however, the farmers showed their power at the polls. After the election of a new governor and a substantial turnover in the legislature, the state government finally met some of the rebels' demands.

Shays's Rebellion was not the only such crisis. In every state, there were large numbers of people, mostly farmers, who needed financial assistance in the form of tax exemption, a moratorium on the payment of debts, or an increase in the currency supply. In an era of economic stagnation and rising taxes, many of those in financial trouble took forceful action to press for change. New Hampshire, Vermont, New Jersey, Pennsylvania, Maryland, Virginia, and South

Daniel Shays and Job Shattuck. These former Revolutionary War soldiers were two leaders of the farmers' insurrection that swept through central and western Massachusetts in 1786–1787. The social tensions that led to Shays's Rebellion were created by the postwar economic crisis. The rebellion underscored the doubts many national political leaders had about the stability of the state governments. (*Source: National Portrait Gallery, Smithsonian Institution, Washington, D.C.*)

Carolina all experienced agrarian unrest in the years 1785–1787. Even though none of these uprisings reached the scale of the Massachusetts protest, they were a clear signal of the economic and social instability of the postrevolutionary era. Massachusetts had barely escaped the collapse of its authority; those in power in other states feared they might not be so fortunate. As George Washington noted, there was no greater evidence "of the want of energy in our governments than these disorders."

The Call for Political Consolidation

Faced with the widespread economic crisis, Washington and other prominent leaders felt growing frustration with the condition of American society. The state governments and thousands of citizens were sinking further into debt, and some debtors were rising up against the authority of the states. These leaders decided that only a strong national authority could deal with the political and economic woes of the individual states. The existing central government was too weak, both politically and structurally, to promote and protect American interests on an international scale. It could scarcely regulate domestic affairs. Shays's Rebellion was simply one striking indication of the need for change.

Throughout the revolutionary era, the American government provided only a loose framework for national unity. In 1777, the Continental Congress had adopted the Articles of Confederation, which gave the central government remarkably little power, especially for a time of war. Under the articles, the congress had the authority to borrow money from foreign nations and to make treaties with them. It lacked, however, the ability to tax American citizens and to draft them into military service. The central government could only request money and manpower from the individual states. The states wielded considerable influence in national decision making. In the Continental Congress, each state had one vote, no matter what its population. Most acts of the congress required a nine-vote majority to pass, and major issues needed unanimous consent. A small coalition of states — or sometimes even a single state — could effectively block collective action when national policies threatened local interests.

The emphasis on the autonomy of the states stemmed from American experience. Before the Revolution, the individual colonies had jealously guarded their right of self-government against increasing incursions by the Crown. After breaking away from royal control, they refused to surrender to another strong and potentially tyrannical central authority — even one of their own making.

Given its inherent (and intended) weakness, the government under the Articles of Confederation did surprisingly well in carrying on the war and making the transition to peace. Each success, however, revealed significant deficiencies. The Continental Congress kept an army in the field long enough to win the war, yet its inability to reinforce, supply, or pay the troops bred low morale and the threat of mutiny, even after the war was over. The congress also established far-sighted policies for the Northwest Territory in 1784–1787. It proposed to settle and develop this huge tract of land north of the Ohio River and east of the Mississippi according to republican principles. The Northwest Ordinances specified the creation of congressionally supervised "territories" (not colonies) that would eventually enter the Confederation as equal states; however, the congress was unable to attain full control over the territories. Although the Treaty of Paris called for the British to evacuate the region, British troops and hostile Indians dominated it until well into the 1790s.

To most critics of the Confederation government, its main weaknesses were not military or diplomatic, but economic. Robert Morris, a wealthy Philadelphia financier who served as Superintendent of Finance in the Confederation government, articulated the concerns of the economic elite. He feared that the nation's meager wealth would disappear if the government did not improve its own access to money. He called for the national government to ally itself with the "interests of moneyed men." Wealthy financiers, both in the United States and abroad, had the capital for private investment and government loans. To encourage their trust, Morris argued for repayment of the government's existing foreign and domestic loans (some of which were owed to Morris and his fellow financiers). He maintained that the national government could not allow individual state governments to determine their own debt-repayment policies. He insisted, rather, that the national government must tax the states to guarantee a source of revenue. Morris and other economic nationalists also believed that the government should regulate interstate commerce. Only by imposing national standards could the United States attain economic stability.

In 1786, delegates from Virginia, Maryland, Delaware, New York, and Pennsylvania (one of whom was Robert Morris) met in Annapolis to discuss interstate commerce. Their deliberations ranged far beyond that topic and resulted in an important proposal: a meeting of all the states should be convened in 1787 to undertake a thorough revision of the Articles of Confederation. The time had come, they felt, to sacrifice states' rights in order to create a nation capable of internal development and worthy of international respect.

Constitution-Making and Compromise

When the fifty-five delegates to the constitutional convention assembled in Philadelphia in May, 1787, only a few of the most famous revolutionary era leaders were absent. Thomas Jefferson and John Adams were serving as American ambassadors in Paris and London, respectively. Adams's cousin Sam, the street leader of Boston's radicals, had not been chosen as a delegate. Patrick Henry of Virginia refused to come, saying he "smelt a rat." Also absent were the political leaders of Rhode Island, who had already opposed Robert Morris's proposal for paying off the national debt. But Morris, George Washington, Alexander Hamilton, and Benjamin Franklin were among the notable revolutionaries who added eminence to the convention. They and their fellow delegates formed a group of highly respected Americans. Most were well-educated, prosperous, and professionally and politically prominent. Most were also committed nationalists and economic conservatives.

One of the youngest but most politically astute of the delegates was James Madison of Virginia. Madison did not dominate debate in the convention; he spent much of his time keeping a meticulous written record of the proceedings. He had done much of his work before the convention, collaborating with his fellow Virginia delegates in drafting a plan for a new national government. This "Virginia Plan," presented to the convention by Virginia's governor, Edmund Randolph, called for a government divided into three parts: a lower house, with representatives elected by the voters of the states; an upper house, with members chosen by members of the lower house; and a national executive and judiciary chosen by the legislature as a whole. The Virginia Plan left some power in the hands of the people, but effectively reduced the power of the states to almost nothing. The central government would have the power to overturn state laws and "legislate in all cases to which the separate States are incompetent, or in which the harmony of the United States may be interrupted."

The clear supremacy of the national government over the states encountered little opposition among the delegates. They did differ, though, on the representation of the states. The Virginia Plan called for proportional representation in the lower house based on population, which was a complete departure from the one state–one vote rule of the Confederation government. Delegates from the less populous states opposed this proposal. They feared that several large states would join together to control the lower house — and thus the upper house and the executive and judiciary as well.

William Paterson of New Jersey presented an alternative "small state" plan. According to Paterson's proposal, each state would have one vote in a unicameral legislature, just as was the case under the Articles of Confederation. Paterson's "New Jersey Plan" also vested broad appointive and administrative powers in the national government, which would have the power to tax the states and to regulate interstate commerce. Like Madison's scheme, the New Jersey Plan was a thoroughly nationalist document.

These two plans engendered intense debate. Eventually the Connecticut delegates devised a third plan, the "Great Compromise," which won wide support. The more populous states won proportional representation in the lower house, and the smaller states were appeased with equal representation in the upper house. Both houses would have to agree on bills before they became law, and the national executive would have limited veto power. This compromise made it impossible for either house or the executive to dominate the government.

Having reconciled the interests of large and small states, the delegates were next confronted with a North-South conflict over slavery. Some northern delegates opposed slavery on moral grounds. Nevertheless, they made important political concessions to southerners. The Constitution did not abolish slavery, nor did it allow Congress to curtail the international slave trade until 1807. In addition, northern delegates consented to count each slave as equivalent to three-fifths of a free person in determining representation in the lower house of Congress. In return, southern states agreed that slaves would be considered as property for determining direct taxes imposed by the national government. The delegates achieved the "three-fifths compromise" by dealing with slavery as a political rather than a moral issue. In framing the Constitution, as in drafting the Declaration of Independence, American political leaders stressed national unity over human rights.

Constitutional Controversy

The proposed Constitution's emphasis on national unity and authority worried many Americans. Three delegates to the convention — Elbridge Gerry of Massachusetts and George Mason and Edmund Randolph of Virginia — refused to sign the document when it was finally completed in September, 1787. Mason later complained: "The very idea of converting what was formerly a confederation to a consolidated government, is totally subversive of every principle which has hitherto governed us. This power is calculated to annihilate totally the state governments." Other prominent leaders who had not attended the convention likewise expressed opposition. Their specific criticisms were varied, but their main concern was, like Mason's, the centralization of power.

The so-called Antifederalists feared that a national government could not govern a vast country in a republican manner. The new constitution vested power in a small group of elected national officials, thus limiting representation and restricting the rights of the individual states and the people themselves. "A representative body, composed principally of respectable yeomanry, is the best possible security to liberty," insisted Melancton Smith of New York. He warned that legislators in the national government would inevitably be drawn from the elite, or "natural aristocracy." Neither the "middling class" nor the poor would be represented. "In so small a number of representatives," Smith added, "there is great danger from corruption and combination."

James Winthrop of Massachusetts argued that the very scope of the national government made it unrepresentative:

> It is the opinion of the ablest writers on the subject, that no extensive empire can be governed upon republican principles, and that such a government will degenerate into despotism. . . . The idea of an uncompounded republic . . . containing six millions of white inhabitants all reduced to the same standard of morals, of habits, and of laws, is in itself an absurdity, and contrary to the whole experience of mankind. The attempt made by Great Britain to introduce such a system, struck us with horror.

Winthrop's reference to Great Britain was calculated to raise the specter of tyranny. A central government could enforce national legislation throughout all the states, he suggested, only with the use of military force.

Proponents of the Constitution responded to the Antifederalist challenge in all the states, but nowhere with more eloquence and sophistication than in New York. There Alexander Hamilton, John Jay, and James Madison wrote a remarkable series of eighty-five essays, collectively titled *The Federalist*. These essays not only answered the Antifederalist attack but advanced a theoretical statement of the new American brand of political science.

James Madison's *Federalist #10* emerged as the most comprehensive and compelling of the essays. Madison challenged the common assumption that only small republics (like the individual states) could guarantee freedom, and he essentially scuttled the basic premise of the Antifederalist argument. He argued that freedom would prosper only in a large republic.

Madison's position reflected various assumptions about human nature and political power. Madison saw — and accepted — a world beset by competition among people motivated by self-interest. It was "sown in the nature of man" that people would form alliances, or factions, to promote their particular interests. Free societies allowed such factions to flourish:

> A landed interest, a manufacturing interest, a mercantile interest, a moneyed interest, with many lesser interests, grow up of necessity in civilized nations, and divide them into different classes, actuated by different sentiments and views.

To eliminate such factions was to curtail freedom or change human nature. The first approach, Madison argued, was undesirable; the second, impossible.

The goal of government was to regulate factions. Small, or minority, factions posed no threat. In a representative government, they could be controlled by the combined power of other minority factions. However, if one faction ever achieved majority status, Madison warned, it could dominate (and therefore oppress) the others. Thus Madison too raised the specter of tyranny: in the hands of a majority faction, representative government could become destructive of the rights and interests of minorities.

Small republics (the individual states) could not guard against such tyranny, Madison argued. They lacked the social and economic diversity that promoted the proliferation of competing factions:

> The smaller the society, the fewer probably will be the distinct parties and interests composing it; the fewer the distinct parties and interests, the more frequently will a majority be found of the same party . . . and the most easily will they concert and execute their plans of oppression.

Greater safety, he concluded, stemmed from greater size:

> Extend the sphere, and you take in a greater variety of parties and interests; you make it less probable that a majority of the whole will have a common motive to invade the rights of other citizens.

Madison's conception of politics differed dramatically from that of his Antifederalist opponents — and from that of earlier Americans. American political leaders had consistently argued that good government depended on good people, whether pious Puritans or virtuous republicans. Madison did not expect that people would be virtuous or self-sacrificing. He expected them to compete with each other in pursuing their own self-interest. His system required only that there be enough competitors to prevent any one interest from achieving dominance. In a sense, he stressed mechanics over morality: the new national government would consist of a multitude of distinct interests, all offsetting one another in a self-regulating system of checks and balances.

The proponents of the Constitution did not rely solely on essay writing to win support. Ratification also required skillful (and sometimes underhanded) political maneuvering. As soon as Congress approved the proposed document, Federalists rushed to hold ratifying conventions before Antifederalists could organize effective opposition. Federalists in Pennsylvania — a powerful coalition led by Philadelphia lawyers and merchants — encountered resistance from several western farming communities that had long resented eastern political dominance. Farmers near more commercially developed towns threw their support behind the Constitution. As one Pennsylvania Antifederalist observed, "the counties nearest the navigation were in favor of it generally; those more remote, in opposition." Pennsylvania, Delaware, and New Jersey ratified the Constitution in December, 1787, and Georgia and Connecticut followed in early 1788.

The ratification battle in Massachusetts loomed as the most critical. Federalists knew that they could not establish an effective national government without Massachusetts. They also knew that the Antifederalists had strong leadership and widespread support there. Elbridge Gerry, one of the state's delegates to the constitutional convention, had refused to sign the finished document, and he urged his fellow citizens to reject it. Powerful political leaders such as Samuel Adams, James Warren, and James Winthrop also opposed ratification. John Hancock, the popular governor, refrained from voicing an official opinion on the Constitution, but his silence suggested sympathy with the Antifederalist position. Moreover, Antifederalist sentiment was intense among former Shaysites and their sympathizers in western farming regions. The only hope for securing ratification, wrote one Massachusetts Federalist, was "to *pack* a Convention whose sense would be different from that of the people."

When the Massachusetts ratifying convention opened in January, 1788, the Antifederalists had a significant majority. Having failed to pack the convention at the outset, Federalists resorted to other methods. Wealthy Federalist delegates used their financial influence to pressure several Antifederalists into changing sides or leaving the convention. Federalist leaders relied on flattery and promises to win the crucial support of John Hancock. Not only did they agree to back him in the upcoming gubernatorial election, they also hinted that if Massachusetts ratified the Constitution and Virginia did not, then he, not George Washington, would become the new nation's first president. Hancock's support and Antifederalist defections gave the convention to the Federalists. They won by a final tally of 187 to 168, the slimmest margin of victory among the first six ratifications.

Equally difficult contests took place in other states, but by the end of June, 1788, ratifying conventions in ten states had given the Constitution the approval it needed to become the supreme law of the land. On July 4, thousands of Philadelphians turned out for a parade in honor of the new Constitution. A band played "The Federal March," specifically written for the occasion. One parade float carried an oversized framed reproduction of the Constitution. Another had a "Grand Federal Edifice" made of thirteen columns, three of which were still unfinished to represent the three states (New York, North Carolina, and Rhode Island) that had not yet ratified the Constitution. By celebrating the new Constitution on the twelfth anniversary of the Declaration of Independence, Philadelphians expressed their pride in the two documents, both written in their city. The parade's timing also symbolized the link between these two most important products of the revolutionary era.

The parade also demonstrated the social reality implicit in Madison's *Federalist #10*. Most of the 5000 people in the procession marched not as individual virtuous republicans but as members of distinct occupational or social groups. Farmers cast seed before them as they walked. On the textile manufacturers' float, weavers operated a loom. Not to be outdone, printers ran a press, reproducing and then distributing a poem that celebrated the Constitution. Behind these floats marched a long line of artisans and professional groups — barbers, hatters, clockmakers, doctors, lawyers, and clergymen. Bringing up the rear were members of Congress and, finally, schoolchildren and teachers. Most blacks, women, and poor whites watched from the sidewalks. All joined together as Americans to cheer the achievements of the revolutionary era just ending. But some Americans remained spectators, both of the parade and of the political process. And those who participated did so with a heightened sense of their respective identities and interests. The United States began its history as a national republic with a society divided by gender, race, class, and economic interests.

9

The Creation of a New Political and Legal Order

In 1787, the Pennsylvania physician and Patriot Dr. Benjamin Rush observed:

> The American war is over, but this is far from being the case with the *American Revolution*. On the contrary, nothing but the first act of the great drama is closed. It remains yet to establish and perfect our new forms of government.

By 1820, American political leaders had met Rush's challenge and devised three "new forms of government."

First, the Federalists led by Alexander Hamilton used the framework provided by the Constitution to create a strong national government. When Hamilton employed governmental powers to assist merchants and to establish a national system of public credit, James Madison and Thomas Jefferson contested his policies. To oppose Hamilton's Federalists, they founded the Democratic-Republican political party. From 1794 to 1816, these two parties struggled for control of the national government; each wanted to use governmental power to assist the social groups that it represented.

Second, state governments quickly and purposefully created a mercantilist (or state-directed) economic system. They established new institutions, such as banks, to provide credit for merchants and traders. To assist farmers and planters, northern and southern legislatures encouraged the construction of turnpikes and canals that would provide cheap and reliable transport. Between 1790 and 1820, state governments created an impressive financial and transportation infrastructure, thereby unleashing a massive expansion of the market economy. The states' policies, like those of Hamilton, favored capitalist entrepreneurs to the detriment of ordinary citizens and prompted political unrest.

Third, American political leaders created new institutions that expanded the "rule of law." Lawyers played an important role in this process; they drafted the new state constitutions and won election to legislative offices in increasing numbers. In addition, the hierarchically organized state judicial systems that enforced criminal statutes and adjudicated civil disputes were devised and staffed by lawyers. Along with the United States Supreme Court, these state courts gave direction to the emergent market economy.

Thus, the years between 1790 and 1820 were among the most creative in the history of the United States. In less than a single generation, leaders devised new institutions that significantly changed the character of American political, economic, and legal life. The "great drama" envisioned by Rush came to pass, but the plot was more complex and troubled than he had supposed. Each of the new institutions was "republican," created by the authority of the sovereign people. Yet wealthy or ambitious Americans exercised considerable control over these organizations, giving them an aristocratic, or elitist, character. Consequently, the creation of a republican polity both unleashed the energy of the American people and revived the social conflicts of the revolutionary era.

A National Fiscal System

The creation of a national system of public credit produced a rapid (and unexpected) division among the framers of the Constitution. Alexander Hamilton of New York and James Madison of Virginia were the main protagonists in this political drama. The son of a wealthy Virginia planter, Madison attended the College of New Jersey (Princeton University) and, between 1784 and 1787, served in the Virginia Assembly. An idealistic, conservative Patriot, Madison was astounded by the "crudeness" of his fellow state legislators who seemed to have only "a particular interest to serve." He tried repeatedly to beat back measures of a "popular cast," such as an "itch for paper money." Madison was unable to prevail on the state level, but did succeed at the constitutional convention in transferring control over many financial issues to the national government. It was his belief that on a national level policies would be determined by "individuals of extended views and of national pride."

Madison got more than he bargained for. As expected, the Electoral College created by the Constitution chose George Washington as the first president of the new government of the United States. A committed nationalist, Washington deliberately chose his cabinet advisors from all regions and the most important states. Thomas Jefferson and Edmund Randolph, both Virginia planters, became secretary of state and attorney general, respectively; Henry Knox of Massachusetts served as secretary of war; and Alexander Hamilton of New York became secretary of the treasury. All were men of "national pride." But Hamilton had an especially purposeful "extended view" of the destiny of the United States. Soon he and Madison were bitter political enemies.

Alexander Hamilton was born on the Caribbean island of Nevis in 1753.

Abandoned by his father, an unsuccessful Scottish merchant, Hamilton was raised by his mother, the daughter of a French Protestant physician. In 1772, he enrolled at Kings College (Columbia University) in New York and joined the Patriot movement. During the war, Hamilton served as a trusted military aide to Washington. He then married into the aristocratic Schuyler family of New York and quickly adopted the social values of the elite. Mobilizing New York landlords, merchants, and creditors, he sought to defend the "security of property" by putting "men in the Legislature whose principles were not of the *leveling kind.*" To guarantee rule by "the rich and well-born," Hamilton advocated an elective monarchy at the constitutional convention.

As secretary of the treasury in Washington's administration, Hamilton used his brilliant talents to enhance the power of wealthy families within American society. His strategy had two interrelated elements. First, Hamilton augmented the power of the national government by creating a new system of public finance. Second, he designed this fiscal system to increase the wealth of the mercantile and creditor classes, thereby winning their support for the new national government.

Hamilton's fiscal system had four parts: redemption, assumption, funding, and credit. As secretary of the treasury, Hamilton submitted legislation in 1791 to redeem the war securities issued by the Continental Congress. He proposed payment of the full face value of the certificates to their present owners. As a congressman from Virginia, Madison opposed Hamilton's scheme as unjust, pointing out that merchants and speculators owned a substantial majority of the $70 million of outstanding Continental securities. These securities had been bought as speculative investments for only a fraction of their original value. Madison proposed that the soldiers, farmers, and investors who had first owned these certificates receive a portion of the redemption fund. The House of Representatives defeated Madison's proposal by a vote of 36 to 13. Some members thought that Madison's plan was too complex. Other congressmen stood to benefit personally from Hamilton's scheme; no fewer than twenty-nine of the sixty-four members of the House held securities themselves.

Madison rose again to oppose Hamilton's "assumption" scheme. The secretary of the treasury wanted the national government to assume and pay off the war debts of the several states. Virginia and several other states opposed assumption. They had already taxed their citizens heavily to retire most of their war debts, and did not want to pay the debts of other states. Moreover, many southerners knew that northern speculators had bought up much of the remaining southern war debt. In fact, forty-seven nonresidents owned $3.2 million of the $8 million in securities still owed by Virginia, North Carolina, and South Carolina. A mere thirteen individuals or firms, mostly located in New York City, controlled nearly $2 million of these securities.

Hamilton made a deft move to win southern support for this monumental financial manipulation. In return for Virginians' support for assumption in the House of Representatives, the secretary of the treasury agreed to move the national capital to a new "federal" district to be laid out along the banks of the Potomac River, between Maryland and Virginia. Despite this accord, many political leaders from the southern states openly questioned the wisdom of Hamilton's financial policies.

Hamilton did nothing to quiet the fears of southerners or northern farmers. He declared his intention to create a permanent national debt. To fund the annual interest on the debt, Hamilton imposed two sets of national taxes. First, he secured legislation increasing the tariffs on imported goods. Once again southern planters protested, for they wanted low-priced British goods. Second, Hamilton placed substantial excise taxes (of up to 25 percent) on whiskey and other domestically produced goods. These internal taxes reminded many Americans of the Stamp Act imposed by Prime Minister Grenville in 1765. They also rekindled doubts concerning the extent of the authority bestowed upon the national government by the Constitution. As Patrick Henry of Virginia had warned at the time, "Your rich snug, fine, fat, Federal offices — The number of collectors of taxes and excises will outnumber any thing from the States." Farmers in western Pennsylvania refused to accept these taxes and forcefully resisted national authority in the Whiskey Rebellion of 1794.

Meanwhile, Hamilton proposed legislation creating the First Bank of the United States — the final part of his fiscal edifice. He hoped that the bank would expand the credit available to merchants and assist the government in the collection of tax revenues. Astounded by the bank bill, Secretary of State Thomas Jefferson joined Madison in active opposition to Hamilton's program. "The incorporation of a Bank," Jefferson argued, was not "delegated to the United States by the Constitution." Reading the Constitution in a "strict" manner, Jefferson pointed out that a national bank was "not among the powers specially enumerated" and was not "*necessary* and proper for carrying into execution the enumerated powers." He therefore urged President Washington to veto the bank bill.

In reply, Hamilton endorsed a "loose" reading of the Constitution. The word "*necessary,*" he told Washington, "often means no more than *needful, requisite . . . useful.*" "If the *end* be . . . within any of the specified powers, and if the measure have an obvious relation to that *end,*" he told the president, "it may safely be deemed to come within the compass of the national authority." Because of his own nationalist sympathies, Washington endorsed Hamilton's position on the bank. Likewise supporting the secretary of the treasury on the issue of excise taxes, Washington dispatched 15,000 United States troops to repress the Whiskey Rebellion.

Backed by the president and a majority of the Congress, Hamilton had created an impressive system of public credit, the revenue measures to fund it, and a national army to uphold it. Many southern planters and backcountry farmers soon realized that Hamilton's financial program favored northern mercantile interests at the expense of their agricultural economies.

TABLE 9.1
Hamilton's Public Credit Policy, 1792

National Government Debt		
Redemption:	Foreign Debt	$11.7 million
	Domestic Debt	42.4 million
Assumption:	State Debts	21.5 million
	TOTAL DEBT	$75.6 million

Annual Revenue		Annual Expenditures	
Customs Duties	$4.4 million	Interest on Debt (at 6%)	$4.6 million
Excise Taxes	0.4 million	Operating Expenses	1.0 million
Miscellaneous	0.6 million		
TOTAL REVENUE	$5.6 million	TOTAL EXPENSES	$5.6 million

Hamilton created a permanent debt to tie merchant capitalists (who owned government securities) to the new Federal regime. He supported customs and excise duties to pay the substantial annual interest on the debt.

The First Party System

Hamilton's policies led to the creation of what historians call the "First Party System." Initially, Madison and Jefferson worked within the existing political institutions to defeat Hamilton's designs. They vigorously pressed their point of view on the president, and they organized an opposition "faction" within Congress. When these measures failed, the two Virginians became political innovators. As the election of 1794 approached, they created a political party, the Democratic-Republicans, to win support among the people. In response, Hamilton's supporters formed the Federalist party.

The social roots of the two parties extended deep into the history of the revolutionary era. Many backcountry farmers who had campaigned for democratic state constitutions after 1776 and who supported the Antifederalist cause in 1787 flocked to the Democratic-Republican banner. And Federalists drew part of their strength from creditors and merchants who led the movement for a strong national Constitution. As a result of the French Revolution, which broke out in 1789, these struggles between political parties took a more ideological form. Chanting the French republicans' slogan of "liberty, equality, and fraternity," American Democratic-Republicans condemned Hamilton's "aristocratic" economic policies and his pro-British foreign policy. For their part, Federalists attacked the "levelling" principles of the French republicans and their repression of Christian churches. Federalists used their control of the national government to give diplomatic support to Britain in its war against the First French Republic. In the Jay Treaty of 1795, Federalists patched up existing disputes with Britain over western lands and maritime rights.

To secure a majority in the national legislature, both parties organized on

TABLE 9.2
Party Outlooks Compared

Federalist	Democratic-Republican
Support for merchants and manufacturers	Aid to artisans and western farmers
Rule by a well-born elite	Rule by an elite of talent
Vote restricted to property-owners	Democratic franchise
A strong judiciary	A strong legislature
A powerful national government	A limited national government
A permanent debt financed by wealthy	Elimination of national debt
Pro-British foreign policy	Pro-French foreign policy

the state and local levels. By 1796, Federalist and Democratic-Republican "legislative caucuses" existed in every state. These caucuses nominated candidates for each congressional district, influenced the selection of senators, and provided institutional support for local politicians.

The first party system dramatically changed the structure of American politics. Before 1794, the political system had been basically "cellular" in nature. Each New England town, or Middle Atlantic and southern county, elected its own officials on the basis of social status or personal abilities. Voters similarly chose local notables to represent their interests in the colonial or state assemblies and in the new national Congress.

After 1794, party affiliation became as important as social prestige in determining the outcome of many elections. "The novel doctrine of new-fashioned republicans . . . that characters should be tried at the bar of public opinion" was "senseless jargon" complained one New England Federalist, who favored the old system of rule by local notables. Nonetheless, Federalists sought to turn ideological principles and party identity to their advantage. "Our combination is to preserve principle," one Federalist asserted, and another party leader accused Democratic-Republicans of seeking "to obtain office, and change the customs and habits of the country." Many voters supported men for office because they stood for a certain set of economic and social policies.

The first party system encompassed the entire society within a single political framework. Federalists confronted Democratic-Republicans in each of the traditional local "cells," breaking down the fragmented character of the old political world. The party system endangered the position of many of the traditional power brokers — the ministers, merchants, planters, and landlords who had connected separate communities to the wider world. Some of these established leaders were politicians by instinct; they found it easy to campaign for office. Many local notables remained committed, emotionally and intellectually, to the older world of deferential politics. They refused to seek the votes of their social "inferiors" or to engage in public debates with the lawyers who became leaders of many local party organizations. Thus, the first party system partially undermined the old system of deferential politics and created a new institutional structure of power and prestige.

This struggle between Federalists and Democratic-Republicans led to a series of major political crises that nearly destroyed the American union. In the election of 1796, Federalist John Adams of Massachusetts succeeded Washington as president. Adams ignored the continued seizure of American merchant ships by Britain yet undertook an undeclared naval war against France in response to similar actions. To suppress domestic criticism of this pro-British foreign policy, the Federalists enacted the Naturalization, Alien, and Sedition Acts of 1798. These acts curbed the voting rights of recent immigrants, many of whom were Democratic-Republicans, and made it illegal to conspire "with intent to oppose any measure or measures of the government of the United States."

Led by Thomas Jefferson, the Democratic-Republicans exploited popular discontent with foreign war and domestic repression. Winning the support of southern planters, western farmers, urban artisans, and ambitious northeastern politicians, Jefferson won election to the presidency in 1800. In a symbolic effort to restore national unity, Jefferson declared in his first inaugural address, "We are all Republicans, We are all Federalists."

The new president's attempt to end the party system was premature. The two parties drew their support from different regions and social groups, and espoused fundamentally divergent policies. On coming into power, the Democratic-Republicans ended internal excise taxes, reduced the size of the army and the national debt, and enacted a new Land Act to encourage rapid settlement of the west. Federalists opposed all these measures, but without success, for Democratic-Republicans held majorities in both houses of Congress.

Jefferson was reelected in 1804, only to face renewed attacks on American merchant vessels by the warring nations of Europe. To force France and, especially, Britain to respect American rights as "neutral traders," the president endorsed legislation imposing an embargo on American shipping. The embargo of 1807 rekindled the fires of party politics. Federalist merchants readily accepted the risks of wartime trade because the potential profits were immense. They assailed the embargo as "unfeeling and odious," a direct attack on the economy of New England and the seaport cities, and they campaigned strenuously to oust the Democratic-Republicans from office.

Despite the Federalists' opposition, James Madison won the presidency in 1808. By leading the nation into war against Great Britain, he triumphed again in 1812. Federalists vigorously opposed the War of 1812. Led by Daniel Webster of New Hampshire, they refused to lend money to the national government or to supply it with troops. It was only in 1815, with the end of the War of 1812 and the French Revolution in Europe, that the political passions generated by the workings of the first party system gradually cooled.

Political parties had not appeared by design. Colonial political leaders had consistently condemned "factions" as dangerous to the safety of the state. The ideology of classical republicanism was hostile to the concept of groups of self-interested politicians; it celebrated "virtuous" leaders who ruled on behalf of the entire people. As a result, neither the state constitutions nor the Constitution of 1787 mentioned parties or sanctioned their appearance. By creating disciplined political organizations, Federalists and Democratic-Republicans deliberately circumvented the checks and balances built into the new republican institutions of government. The inherently antirepublican character of bitterly contending parties alarmed most politically minded Americans. "If I could not go to heaven but with a party," Jefferson declared, "I would not go there at all."

On the other hand, parties were the logical result of another aspect of republican thought, the doctrine of popular sovereignty. Ultimate power in the new American republic lay in the hands of the people, and parties were the only way that the populace could be mobilized for effective action. American politics had not yet become "democratic," since most Federalist and Democratic-Republican leaders assumed that the people would defer to their views. However, the first party system partially undermined the elitist-controlled political system established by the Constitution. It created a new institutional structure that arrayed contending factions of elites against one another and forced them to appeal for popular support. At the same time, party organizations united the ordinary citizens of the American republic and provided them with a conscious sense of membership in the political order and a measure of electoral power.

The Structure of Politics

The battles over Hamilton's fiscal system and Jefferson's foreign policy took place in a new national political arena. Between 1775 and 1789, the states had been the units of representation in the Confederation government. To exert pressure on the national level, interest groups had to influence a state's delegation to the federal congress. As a result, the state legislatures controlled politics both within their own boundaries and for the nation as a whole.

The Constitution dramatically changed this structure of power. Property-owning white adult males in every state voted, either directly or indirectly, for members of the Electoral College. In turn, these electors selected the president of the United States. More important, voters in each congressional district directly elected a member of the House of Representatives, the lower house of the national legislature. State legislatures still retained a significant role in the selection of the central government. Some legislatures chose members of the Electoral College and (until the passage of the Seventeenth Amendment to the Constitution in 1912) all of them elected members of the United States Senate. In spite of this, the Constitution did shift the locus of power away from the state legislatures. "The powers given to the federal body for imposing internal taxation will necessarily destroy the state sovereignties," William Findley of Penn-

Election Day in Philadelphia, 1816. The first party system changed the character of American politics. Elected officials were still well-dressed gentlemen, but ordinary artisans (such as the orator in the worker's apron at the far left) began to play a more active role in political affairs. (*Source: Historical Society of Pennsylvania.*)

sylvania prophesied in 1788, "for there cannot exist two independent sovereign taxing powers in the same community, and the strongest will, of course, annihilate the weaker."

The imposition of excise taxes and the vigorous suppression of the Whiskey Rebellion of 1794 demonstrated the immense potential power of the national government. The war crisis with France in 1798 pointed out the ability of the national government to quell ordinary political dissent as well. Using the federal courts, the Adams administration indicted dozens of Democratic-Republicans for seditious libel. Federalist judges assisted prosecutors in obtaining guilty verdicts and sentenced numerous Democratic-Republican newspaper editors to lengthy jail terms. Jefferson and many other Democratic-Republicans thought the New England Federalists were using the national government to rule in an arbitrary and tyrannical manner. "We are completely under the saddle of Massachusetts and Connecticut," Jefferson wrote to his friend John Taylor, "[and] they ride us very hard, cruelly insulting our feelings as well as exhausting our strength and substance."

Jefferson and Madison anonymously drafted resolutions for passage by the Kentucky and Virginia legislatures. These resolutions attacked the Sedition Acts as contrary to the First Amendment to the Constitution, which protected the rights of freedom of speech and freedom of the press. Equally important, the resolutions claimed substantial sovereign powers for the states. The limits of the powers of the national government had already been clarified in the ten amendments to the Constitution that had been ratified in 1791. Proposed in response to Antifederalist criticisms, these amendments protected the rights of citizens; the Tenth Amendment declared that "powers not delegated to the United States by the Constitution, nor prohibited by it to the States, are reserved to the States respectively, or to the people."

Jefferson went further than that in the Kentucky Resolutions; he asserted that the national government was a "compact" constituted by the "several states" and that "whensoever the General Government assumes undelegated powers, its acts are unauthoritative, void, and are of no force." In the Virginia Resolutions, Madison asked other states to concur with the Virginia legislature "in declaring, as it does hereby declare, that [such] acts . . . are unconstitutional." The political intent of these resolutions was clear. Madison and Jefferson were using their influence in the states controlled by the Democratic-Republicans to contest Federalist authority on the national level.

The Federalists adopted a similar strategy during the administrations of Jefferson and Madison. Jefferson's view of America's destiny was very different from that of Hamilton. He dreamed of a society filled with families of yeoman farmers and independent artisans. To open new lands for southern tenant farmers and hard-pressed New England freeholders, Jefferson advocated the rapid settlement of the west. He drafted the Land Ordinance of 1784, which called for the creation of new western states and banned slavery within their bounds.

As president, Jefferson pursued this vision of an expanding commercial agricultural society. Early in his administration, the Spanish government first threatened to restrict American use of the Mississippi River and next secretly transferred the vast territory of Louisiana to France. Jefferson wrote to a friend:

> There is on the globe one single spot, the possessor of which is our natural and habitual enemy. It is New Orleans, through which the produce of three-eighths of our territory must pass to market, and from its fertility it will ere long yield more than half of our whole produce and contain more than half of our inhabitants.

In 1803, Jefferson purchased New Orleans and the entire territory of Louisiana from the French government, at the time headed by Napoleon I. By this magnificent diplomatic coup, the president moved the western boundary of the United States from the Mississippi River to the Rocky Mountains, nearly doubling the nation's land area.

The legality of the Louisiana Purchase was as doubtful as that of the creation of the First Bank of the United States and the Sedition Acts, for the Constitution did not specifically provide for the acquisition of new territories. The political implications of this acquisition were equally controversial, since it threatened the eastern-oriented Federalist party with the permanent loss of power. "We rush like a comet into infinite space," warned Fisher Ames, a prominent New England Federalist, on learning of the purchase. "The western states, multiplied in number, and augmented in population," predicted Federalist Harrison Gray Otis, "will control the interests of the whole." To avoid domination by southern and western Democratic-Republicans, Massachusetts Federalist Timothy Pickering advocated separation from "the *aristocratic democrats* of the South" and the breakup of the Union.

The embargo of 1807 and the War of 1812 gave new force to the separatist sentiments that were prevalent in New England. "We have always been led to believe that a separation of the States would be a great evil," the editor of the Newburyport, Massachusetts *Centinel* observed in 1813, "[and] we still think it an evil." "But rather than prosecute the present war . . .," he continued, "we think it by far the least of two evils." Like Madison and Jefferson in 1798, these Federalists argued that the United States was simply a collection of "several independent confederated republics."

To defend the interests of their party, states, and region, New England Federalists called the Hartford Convention in 1815. Resisting the advice of Pickering and other secessionists, the convention endorsed seven constitutional amendments. These amendments would give the New England states an effective veto over certain important issues; an affirmative vote by two-thirds of the states would be required before the national government could impose an embargo, declare an offensive war, or admit new states into the Union. Even as it sought compromise, the convention declared that in cases of "deliberate, dangerous" violations of state sovereignty or popular liberties, "it is not only the right but the duty of such a state to interpose its authority for their protection."

In 1800, Jefferson's victory had ended the threat of southern secession from the Union. In 1815, the conclusion of a peace treaty with Great Britain quashed the plans of separatists in New England. Yet these threats to the Union were serious, for they showed that many Americans had not accepted the national structure of politics created by the Constitution in 1787. Accustomed to the primacy of local and state politics, they refused to accept the authority of the president and Congress when it seriously affected their interests.

The legacy of these early confrontations between state and national sovereignty was twofold. After 1815, most American political leaders advocated a very limited role for the national government in order to avoid similar constitutional controversies. Where national action was necessary, such as in the origination of tariff policies or the admission of new states into the Union, politicians carefully balanced the interests of the various regions and of powerful social groups. These cautious strategies preserved the Union until the conflict over the expansion of slavery prompted the secession of most southern states. Then, in the Civil War, the paramount authority of the national government was confirmed by force of arms.

State Mercantilism

Beginning in the 1780s, American governments — national, state, and local — systematically promoted economic development. This effort resembled the mercantilism of European nations in that it involved the use of political authority to mobilize the society's resources. Both republican ideology and economic necessity justified this political intervention. "The establishment of useful manufactures is closely connected with the public weal," the New York legislature declared in 1790, as it granted a loan to a pottery factory. Mercantilist assumptions determined American economic policy for the next half-century. Chief Justice Black of Pennsylvania declared in a court decision in 1853:

> It is a grave error to suppose that the duty of a state stops with the establishment of those institutions which are necessary to the existence of government. . . . To aid, encourage, and stimulate commerce, domestic and foreign, is a duty of the sovereign, as plain and universally recognized as any other.

Aid to business enterprises took various forms. Following the ratification of the Constitution, the United States government embarked on a program of economic nationalism. The Tonnage Act, passed by Congress in 1789, closely resembled the British Trade and Navigation Acts. To encourage the American shipping industry, the Tonnage Act restricted the transport of goods between American ports to vessels owned by United States citizens. In addition, it levied a special duty of 50 cents a ton on goods imported in foreign-built or foreign-owned ships.

Congress also passed a series of tariff acts that placed duties on imported goods. Like the despised Townshend Duties of 1767, these acts were designed primarily to raise revenue. Unlike that earlier British legislation, they did so with great success. Tariffs provided two-thirds of the income of the national government between 1790 and 1860. In addition, tariff legislation assisted the growth of American industry. By 1794, import duties on shoes and various other foreign manufactured items reached 20 percent of the goods' value. These protective duties subsidized domestic manufacturers (at the public's expense) by raising the price of imported goods.

State governments fostered economic development with even greater dedication. In the thirty years from 1789 to 1819, state legislatures issued more than 2500 charters to business corporations. These charters gave corporations a defined legal existence, enabling them to enter into contracts and to attract capital from private investors. At first, legislatures granted corporate charters selectively, directing private resources primarily into transportation projects. Of the 326 charters issued before 1800, 207 (or 63 percent) were for the construction of turnpikes, bridges, and canals. By 1812, the Commonwealth of Pennsylvania had chartered 55 turnpike companies, New York had granted articles of incorporation to 57 such companies, and Massachusetts to 105. In many cases, these charters guaranteed monopoly rights over certain routes in order to encourage an adequate level of investment.

Governmental participation in the construction of canals was much more extensive than the mere granting of legal privileges. The amount of capital required to build a canal was enormous. Only in eastern New England, with its densely concentrated population, was the prospect of profit imminent enough to attract a sufficient number of private investors. Elsewhere, state governments took the initiative, concluding that canal building was a worthwhile endeavor in the long term because it would open up interior regions for settlement and for trade. Pressed by George Washington and other influential planters, the Maryland and Virginia legislatures created the Potomac River Canal Company in 1785 and appropriated state funds for its endeavor. Subsequently, other states poured massive amounts of public funds into canal projects. Of the total investment of $132 million in the canals of the trans-Appalachian region between 1815 and 1860, state governments arranged for $117 million through grants, bond issues, or taxation. This use of public financing permitted the rapid settlement of the states of the Ohio River Valley.

The Commonwealth of Pennsylvania took the most active role of all the states in assisting economic development. In 1809, its legislature granted a loan to private individuals to construct a steel mill. Other ventures outside the field of transportation quickly followed. Arguing simply that "works of public importance deserve public encouragement," the commonwealth invested in a variety of mixed public and private enterprises. By 1844, the state government was part-owner of more than 150 companies.

Purposeful government intervention gave public direction to American economic development. In addition, government-financed projects increased the rate of investment, of capital accumulation, and of western settlement. The Erie Canal, financed and constructed by the state government of New York, was the crowning achievement of state mercantilism. The building of this magnificent canal, connecting the Hudson River and Lake Erie, required an immense importation of laborers (thousands of Irish immigrants) and capital resources from Europe. The end of the Napoleonic Wars had dramatically reduced the borrowing needs of European governments and created, as Governor Clinton of New York correctly discerned, "a vast disposable capital in Great Britain." New York pledged public taxes to guarantee its canal bonds, and British investors poured millions of dollars into the project. (See map on p. 210.)

By the time of its completion in 1825, the Erie Canal had already helped to make New York City the largest and wealthiest American city. Most trade with the Great Lakes' states of Ohio, Indiana, and Illinois moved along the canal, as did tens of thousands of bushels of wheat grown on the fertile lands between Albany and Buffalo. No fewer than 900,000 settlers already lived in these upstate counties of New York. Most of them had migrated from New England, where two centuries of rapid population growth had diminished the size of farms and depleted the fertility of the soil. Expansion into the interior of North America gave these farmers — and the society as a whole — access to new and more productive natural resources. Thus, the construction of canals, in conjunction with other state mercantilist activities, spurred the rapid economic development of the new American republic.

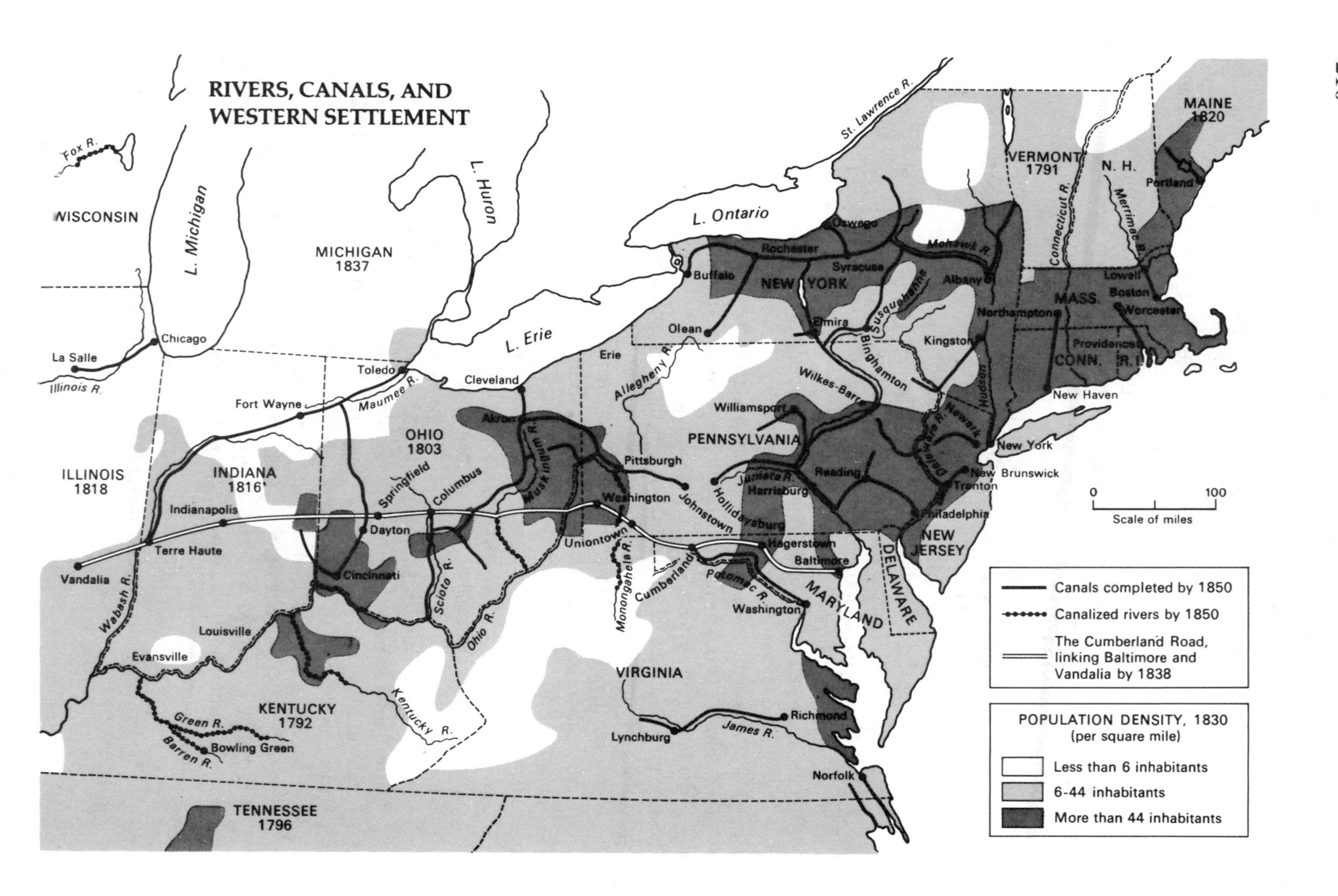
RIVERS, CANALS, AND WESTERN SETTLEMENT
Canals completed by 1850
Canalized rivers by 1850
The Cumberland Road, linking Baltimore and Vandalia by 1838
POPULATION DENSITY, 1830 (per square mile)
Less than 6 inhabitants
6-44 inhabitants
More than 44 inhabitants
0
100
Scale of miles
MAINE 1820
N. H.
VERMONT 1791
MASS.
CONN.
R.I.
NEW YORK
PENNSYLVANIA
NEW JERSEY
DELAWARE
MARYLAND
VIRGINIA
OHIO 1803
INDIANA 1816
ILLINOIS 1818
MICHIGAN 1837
WISCONSIN
KENTUCKY 1792
TENNESSEE 1796
L. Michigan
L. Huron
L. Erie
L. Ontario
St. Lawrence R.
Merrimac R.
Connecticut R.
Hudson
Mohawk R.
Susquehanna
Delaware R.
Juniata R.
Potomac R.
Allegheny R.
Monongahela R.
Muskingum R.
Scioto R.
Ohio R.
Kentucky R.
Maumee R.
Wabash R.
Illinois R.
Fox R.
Green R.
Barren R.
James R.
Portland
Lowell
Boston
Worcester
Providence
New Haven
Northampton
New York
New Brunswick
Newark
Trenton
Philadelphia
Kingston
Albany
Binghamton
Oswego
Syracuse
Rochester
Elmira
Buffalo
Olean
Wilkes-Barre
Reading
Harrisburg
Williamsport
Hollidaysburg
Johnstown
Pittsburgh
Erie
Washington
Hagerstown
Baltimore
Cumberland
Uniontown
Richmond
Norfolk
Lynchburg
Cleveland
Akron
Columbus
Springfield
Dayton
Cincinnati
Toledo
Fort Wayne
Indianapolis
Terre Haute
Vandalia
Louisville
Evansville
Bowling Green
Chicago
La Salle

Mercantile Capitalism

Even as the state governments used public taxes to attract capital from foreign investors, they encouraged the creation of an extensive privately owned domestic banking system. Before 1776, there were no banks in America. Merchants formed business partnerships or obtained private loans to finance commercial or manufacturing enterprises. During the War of Independence, leading merchants transformed these informal credit arrangements into formal banking institutions. The president of the Bank of North America wrote from Philadelphia in 1784:

> When the bank was first opened here, we established our books on a simple mercantile plan, and this mode . . . pointed out by experience alone, has carried us through so far without a material loss or even mistake of any consequence.

By 1791, when Congress established the First Bank of the United States, state-chartered banks existed in New York, Massachusetts, Maryland, and Pennsylvania. Merchants controlled most of these institutions, using them primarily as a source of short-term credit to pay for imported goods. By 1795, there were twenty-one banks in the United States, and by 1812, there were ninety. By that time, state-chartered banks held more than $15 million in gold and silver. In addition, they had more than $52 million out in loans or circulating as negotiable bills of credit. The First Bank of the United States provided monetary stability for this private banking system. It accepted bank notes in payment of tariffs and then presented them for redemption in specie, forcing the state-chartered banks to maintain adequate reserves.

The existence of this complex banking system indicated that the United States had entered a new stage in its financial development. Money, markets, and rational economic behavior were all to be found in America prior to 1776, but the real center of financial power and decision making lay in England and Scotland. Colonial merchants relied on British sources of credit and lacked the institutions to mobilize capital for noncommercial ventures. Partly as a result of the weakness of the American credit system, a highly developed market economy existed only in the seaport cities and among tidewater planters.

By 1800, however, business corporations, complex credit institutions, and new forms of economic enterprise were playing a more extensive role in American life. Again, governments took the lead. To encourage technological innovation, the United States government created the Patent Office. By registering an invention with the office, a citizen could obtain the exclusive right to manufacture, sell, or use it. This establishing of a government-protected property

Settlers wanted farmland that was near navigable rivers and canals; this demand swelled population density in those areas. Transportation of goods by water was so efficient and cheap that, before 1850, railroads played a very small role in American economic development.

right fostered an outpouring of mechanical ingenuity; the number of patents increased from an average of 20 per year between 1790 and 1795 to 200 per year between 1808 and 1824.

Early nineteenth-century American merchants began developing new types of financial institutions. Around 1800, the merchant house of Alexander Brown and Sons of Baltimore cut back its trade in commodities, such as wheat exports and textile imports. Instead, the Browns started to specialize in "paper" transactions, becoming financial brokers. They charged merchants a fee of 2 percent for buying bills of exchange drawn on foreign banks; then they redeemed the bills in Europe or used them to speculate in the international money market. In Philadelphia, the firm of S. and M. Allen and Sons provided another specialized financial service. The Allens organized a system of state-authorized lotteries to raise capital for public institutions, such as schools, and for private transportation and manufacturing ventures. Because they tapped the small savings of thousands of Americans, lotteries were both popular and successful. The number of offices run by the Allens grew from three in 1809 to sixty in 1827.

Previously, small merchant firms had stood at the center of the American commercial system. Such firms handled a wide array of business — trade, insurance, money lending — as the situation demanded. These diversified merchant entrepreneurs were pushed off the center of the economic stage by a wide variety of specialized bankers, brokers, and commodity traders. These financial specialists had less range and versatility than eighteenth-century merchants, but were no less audacious. Moreover, firms like those of the Browns and Allens became more accomplished practitioners of their trades because they devoted all their energies to a narrow field. With this division of labor, the American financial system advanced to a more efficient level.

The overall effect of the new system of banks, financial institutions, and state mercantilist enterprises was the encouragement of a sustained process of economic growth. Public corporations attracted massive amounts of foreign capital for large-scale transportation projects. By the 1820s, Europeans had invested \$80–90 million in the United States, and this amount reached \$200–250 million by the 1830s. This foreign capital permitted the rapid settlement of the trans-Appalachian west. At the same time, the new system of domestic banking provided short-term credit for merchants and amassed funds for the expansion of the shoe, textile, ship-building, and iron industries.

This accumulation of foreign and domestic capital promoted a substantial increase in national prosperity. The per capita wealth of American citizens had returned to its prewar level by 1790. After that, the rapid growth of the population (from 3.9 million in 1790 to 9.6 million in 1820) strained productive capacity, resulting in a slow growth in per capita wealth. Finally, in the 1820s and 1830s, the positive effects of a generation of institutional development and capital accumulation became apparent. Per capita wealth soared from an index figure of 67 in 1820 to 100 in 1840. (See Table 9.3.)

Although government intervention expanded the total wealth of American

TABLE 9.3
Economic Growth and Social Stratification

Year	Population (millions)	Nonfarm Labor Force	Per Capita Wealth*	Wealth Owned by Top 10% (estimated)
1800	5.3	17.4%	64.4	45%
1820	9.6	21.0%	67.7	50%
1840	17.1	36.6%	100.0	55%
1860	31.4	46.8%	137.0	60%

SOURCE: Paul A. David, "The Growth of Real Product in the United States Before 1840," *JEH* 27 (1967), Tables 4 and 8; and Jeffrey G. Williamson and Peter H. Lindert, *Long-Term Trends in American Wealth Inequality* (University of Wisconsin Press, 1977).

*Based on an index in which 1840 = 100.0.

The expansion of the nonagricultural labor force after 1820 brought about a rapid growth in per capita wealth and increasing social inequality. By 1860, the top 10 percent of American families owned no less than 60 percent of the nation's wealth.

society, it also increased the extent of economic inequality. In 1775 and in 1790, the top 10 percent of the wealthholders in New England and the Middle Atlantic states owned about 50 percent of the society's resources; by 1840, their share had increased to 70 percent. Economic development had greatly augmented the fortunes of the wealthier citizens: the merchants, manufacturers, and capitalist financiers. Farmers, artisans, and workers enjoyed much more modest rises in their standards of living. By 1840, their relative position in the economic order had worsened significantly.

Thus, the American system of state mercantilism was not socially neutral. Legislatures directly rewarded innovative entrepreneurs and achievement-oriented capitalists by giving them special charters, funneling public money through their hands, and perfecting a legal framework designed to suit their needs. In many cases, the managers of public enterprises were also private investors in the same companies; they acted to promote their personal interests as much as the common good. State mercantilism obtained gains in the gross national product at the cost of a growing inequality in the distribution of wealth.

Most public leaders defended these policies as contributing to the "commonwealth" of the society — which they did. Some contemporaries took a more critical view, however. They argued that state mercantilism, and especially the creation of privileged capitalist corporations, undermined the republican principles of liberty and equality for which Americans had fought in 1776. Pointing to the growing inequality of wealth, these critics warned of the disparities in social status and political power that would inevitably follow. "In this country," a Pennsylvanian argued, "sovereignty is vested in the people themselves, and whatever power is granted to corporations, is so much abstracted from the people themselves." This criticism of state mercantilism had as much validity as the argument in its defence. In the name of economic development, the system of state governmental assistance to private business enterprise had enhanced the power of merchants, manufacturers, and other wealthy Americans.

Lawyers and the Rule of Law

In the twentieth century, wars for national independence have often produced authoritarian political regimes. In some new nations, charismatic leaders rely on their personal authority to rule in an arbitrary fashion. Elsewhere, military regimes use armed force to suppress political dissent. In still other new nations, privileged social groups mask their authoritarian rule behind the facade of representative institutions.

The American War of Independence yielded a different result. The governments established during the war adhered, in most respects, to the "rule of law." Authoritarian institutions existed only in the southern states, where white slaveowners continued to control the lives of blacks in an arbitrary and essentially lawless manner. Otherwise, the American people repudiated monarchical and other nonrepresentative forms of authority. In 1812, Chief Justice Theophilus Parsons of the Massachusetts Supreme Court declared in *Ainslie v. Martin:*

> This people . . . considered the several aggressions of their sovereign on their essential rights as amounting to an abdication of his sovereignty. The throne was then vacant; but the people, in their political character, did not look after another family to reign; nor did they establish a new dynasty; but assumed to themselves, as a nation, the sovereign power. . . . Thus the government became a republic.

This fortuitous result reflected the traditional predominance of legally defined political procedures. In the half-century before 1776, the colonists had created broadly based representative assemblies that routinely enacted new laws and amended old ones. Long before they became republics in name, American governments were republican in fact.

In the colonial period, the American settlers had also gradually developed court systems to adjudicate criminal, personal, and economic disputes. Merchants, planters, and land speculators frequently used these courts to enforce written contracts, extract payment from debtors, and uphold property titles. At times, backcountry farmers, impoverished tenants, and hard-pressed artisans assailed this legal regime as unjust and oppressive. "There should be no Lawyers in the province," proclaimed angry North Carolina Regulators, and debtors in New Jersey attacked lawyers as "Private Leaches, sucking out our very Hearts Blood." These protests were to no avail. The court system, along with the legal profession that staffed it, became an ever more important part of colonial life.

The emergence of an influential American bar paralleled the evolution of a legally oriented political culture. "How greatly elevated, above common People, and above Divines is this Lawyer," young John Adams noted after listening to a leading member of his profession. Lawyers commanded impressive fees and pursued elegant life-styles. John Tabor Kempe, who became attorney-general of New York in 1759, realized a yearly income of £880 from his private law practice, and William Livingston averaged nearly £700 a year during his twenty-four years at the New York bar. Lawyers' incomes were more modest in country areas, where the volume of lucrative commercial transactions was lower. Yet Waightsill Avery, a backcountry Virginia and North Carolina lawyer, received £164 in fees in one six-month period, making him one of the more affluent residents of this newly settled region.

To solidify their wealth and prestige, established lawyers challenged the competence of untrained "pettifoggers." In Virginia, the House of Burgesses enacted legislation requiring that men "learned in the law" license new practitioners. New York lawyers formed a bar association in 1748; by the 1760s, this association demanded that a prospective lawyer either serve a five-year apprenticeship or attend college and then work as a clerk in a law office for three years. Subsequently, Boston lawyers set a hefty minimum apprenticeship fee of £100, intended both to increase their incomes and to keep the profession small and exclusive.

Lawyers exercised their newfound authority during the Revolution. At state constitutional conventions, they translated general republican principles into precise legal language. The resulting documents often reflected the concerns or enhanced the authority of the legal profession. The Massachusetts Constitution of 1780, for example, contained a clause that gave the general court "full power and authority . . . to make, ordain, and establish . . . reasonable orders, laws, statutes, and ordinances." Although legislative supremacy was a key republican principle, the doctrine of judicial review was more problematic. Yet the above clause also declared that laws must not be "repugnant or contrary to this Constitution," thus raising the prospect that courts might review (and reject) legislative enactments.

The decision to consider both constitutions and legislation as "law" created a single, but at the same time complex, web of political and legal authority. To further complicate things, eleven of the thirteen states followed the example of the New Jersey Constitution of 1776 in declaring that "common law of England . . . as . . . practiced in the Colony, shall still remain in force." Only those trained in law could sort out the complexities in the three sources of authority: constitutional provisions, legislative statutes, and common law precedents. The United States Constitution of 1787 added yet another dimension to the intellectual problem of the location of legal authority by specifying the ultimate primacy of national law. The so-called supremacy clause declared that the

> . . . Constitution, and the laws of the United States which shall be made in pursuance thereof . . . shall be the supreme law of the land: and the Judges in every State shall be bound thereby, anything in the Constitution or laws of any State to the contrary notwithstanding.

The rule of law was a simple concept, but its implementation would have intricate and far-reaching consequences.

The rule of law formed the basis for an increase in the number and the power of lawyers. There were 24 lawyers for every 10,000 residents in Massachusetts in 1775 and in 1790, but 70 lawyers per 10,000 by 1810 and 87 by 1820. In South Carolina, the proportion of lawyers to residents doubled from 19 per 10,000 in 1771 to 40 per 10,000 by 1820. The legal profession became the most important single source of political leaders, especially for the higher levels of government. Of the 764 lawyers in Massachusetts before 1810, no fewer than 338 (or 44 percent) won election to office. By 1820, even though lawyers made up less than 1 percent of the adult male population, they constituted 15 percent of the representatives in the Massachusetts legislature and 35 percent of the senators. Even more striking, 63 percent of all United States congressmen from Massachusetts between 1789 and 1839 were trained as lawyers.

The growing political power of the legal profession alarmed both backcountry farmers and wealthy gentlemen. Amos Singletary, an Antifederalist farmer in Sutton, Massachusetts, warned that lawyers supporting ratification of the Constitution of 1787 "expect to get into Congress themselves . . . and get all the power and money into their own hands, and then they will swallow up all us little folk, like the great leviathen." Using the elevated language of republican political theory, Antifederalist Dr. Nathaniel Ames (brother of archconservative Federalist Fisher Ames) arrived at a similar conclusion. "The making of lawyers [into] legislators," he argued in 1799, "seems to defeat the grand principle of keeping the Legislative Department distinct and separate from the Judiciary."

In various states, popular distrust of lawyers had led to the enactment of laws to diminish the influence of the legal profession. In Massachusetts, an "Act for Rendering Processes in Law Less Expensive" placed most court actions initially in the hands of local justices of the peace, and a "Referee Act" encouraged the settlement of civil suits through arbitration. Reflecting popular sentiment, many Jeffersonian Democratic-Republicans supported efforts to simplify the common law and to reduce educational requirements for the practice of law.

The attack on the elitism of the legal profession drew support from upwardly mobile individuals as well as from democratic reformers. Over the years, the legal profession had become an exclusive and self-perpetuating caste. Only wealthy families could afford to send their sons to college to qualify them for the bar, and the costly apprenticeship system similarly discriminated against those from the middle and lower ranks of society. To correct this social bias, most states reduced the educational standards for admittance to the bar. By the 1820s, the composition of the American legal profession was more democratic than ever before.

In *Democracy in America* (1835), however, Alexis de Tocqueville described the legal profession as "the most powerful existing security against the excesses of democracy." The French aristocrat's conclusion was somewhat flawed, for he associated primarily with established Federalist lawyers during his visit to America and did not fully appreciate the new diversity of the profession. But Tocqueville correctly discerned that the American judicial system had become a hierarchical and authoritarian institution. Until the early nineteenth century, for example, juries of ordinary men decided questions of *law* as well as of *fact*. Then, rather suddenly, judges assumed the authority to tell juries which law to apply in a given case and how to interpret the law. If a jury brought in a verdict that did not follow that interpretation, a judge would often declare the verdict "contrary to the law" and therefore void. This judicial aggressiveness was in part due to the publication of decisions rendered on appealed cases by the supreme courts of the various states. By "declaring" the law, these printed decisions promoted uniformity throughout the court system. The influential work *Commentaries on American Law* (4 volumes, 1826–1830) by Chancellor James Kent of New York similarly imparted a new degree of predictability to court decisions, as local judges and lawyers argued and decided cases on the basis of precedents cited by the eminent jurist.

Democratic reformers contested the centralization of power in the hands of judges and especially their reliance on common law principles. These reformers demanded the codification of the common law in legislative statutes to ensure the primacy of republican political institutions. As Ohio lawyer John Milton Goodenow argued in the *The Principles and Maxims of American Jurisprudence* (1819), "the judge . . . is *governed* himself by *positive* law, and executes and enforces the will of the supreme power, which is the will of THE PEOPLE." Nonetheless, the judicial system had created a legal elite, which had the power to influence American social and economic development.

Law and Economic Process

Judges recognized that law was less a sacred body of immutable truths than the product of political power and will. Just as the state governments enacted mercantilist legislation to assist economic development, judges increasingly interpreted statutes and common law rules in an "instrumentalist" fashion. If courts strictly adhered to precedent, a judge of the New York Supreme Court declared in 1799, "we must hope of little improvement in our commercial code. A single decision, though founded on mistake, would become of binding force, and by repetition, error might be . . . heaped on error." Therefore, "we are at liberty," another New York high court judge argued in 1802, "to adopt such a construction as shall most subserve the solid interests of this growing country."

John Marshall, Chief Justice of the United States Supreme Court from 1801 to 1835, embraced judicial activism and legal instrumentalism. As a Federalist, he gave these doctrines a conservative and nationalistic twist. In *Marbury v. Madison* (1803), Marshall declared that the Supreme Court could decide the constitutionality of congressional legislation. Two decades later, in *Cohens v. Virginia* (1821), the chief justice ruled that verdicts of state courts could be appealed to the Supreme Court. These decisions were indicative of both the increasing centralization of the judicial system and its role as the arbiter of constitutional legitimacy.

Marshall's conservatism became evident in a series of decisions regarding property rights. Most state constitutions contained "Bills of Rights" that protected the property, religious beliefs, and personal liberty of their citizens from arbitrary governmental actions. The Fifth Amendment to the Constitution similarly prohibited the national government from depriving citizens of their liberty or property "without due process" of law. These guarantees were both procedural and substantive. In procedural terms, the court system resolved disputes in a legal forum that was fairly free from political interference and private passion. And certain substantive rights were established, which even representative governments could not infringe.

Marshall interpreted these constitutional guarantees to give maximum protection to existing (or "vested") property rights. Article I, Section 10, of the Constitution barred the states from enacting any "law impairing the obligation of contracts." In the case of *Fletcher v. Peck* (1810), Marshall ruled that a land grant made by the Georgia legislature constituted such a contract. Therefore, he asserted, the original grant could not be cancelled by a subsequent legislature, even though it had been obtained through fraud and bribery. Similarly, in *Dartmouth College v. Woodward* (1819), Marshall declared that the royal charter of 1769 gave the college "corporate powers and privileges." Hence, he ruled that the New Hampshire legislature could not change the charter to make the college into a public university.

The *Dartmouth College v. Woodward* decision had far-reaching economic implications. By equating charters with contracts, Marshall extended the contract clause of the Constitution so as to protect corporations as well as individuals. As Chancellor Kent pointed out, the decision threw "an impregnable barrier around all rights and franchises derived from the grant of government." Using this decision, turnpike, canal, and banking corporations avoided regulation by the state governments that originally had chartered them. "The government must control these institutions," the attorney for the state of New Hampshire warned, "or they shall control the government."

Judges with republican political philosophies heeded this warning. On the whole, they were less concerned with protecting vested property rights than encouraging economic development. In 1795, for example, the Massachusetts legislature enacted the Mill Dam Act. This legislation allowed owners of grain and textile mills to flood adjacent farming lands if they paid an annual rent. Although the act deprived farmers of the use of their lands without their consent and "without due process of law," it was upheld by the Massachusetts Supreme Judicial Court. In *Stowell v. Flagg* (1814), the court barred farmers from using normal legal procedures, such as a suit at law for trespass, to reclaim the use of their flooded property. It was "the design of the legislature . . . to favor the owners of mills," Chief Justice Parker proclaimed, "as the law is, so we must declare it."

These judges followed legal philosophies that contrasted sharply with Marshall's support for vested property rights. In *Palmer v. Mulligan* (1805), a New York mill-owner attempted to use common law principles (which gave priority to the first user of a resource) to prevent the construction of another mill on the same river. Denying his petition, the New York Supreme Court declared that monopolies based on doctrines of prior use were contrary to the "public" interest, "which always [benefits from] . . . competition and rivalry."

The conflict among the legal philosophies of various influential judges indicated the central importance of the legal system in the American economy. State governments intervened continuously in economic affairs, enacting legislation that favored some social groups at the expense of others. Lawyers and judges acted similarly, protecting the economic rights of their clients and giving their own definitions to rights of property. What they claimed was the public interest was in fact a fragmented and competing set of private rights.

The evolution of contract law clearly showed the increasing primacy of individual goals in legal doctrine and in economic reality. In the 1790s, some lawyers and judges, such as Francis Dana and Theophilus Parsons of Massachusetts, argued that "the existing rights of the public" should not be infringed by private contracts. In fact, some jurists of the time still supported the older "equitable" theory of contract, which permitted courts to intervene to ensure a fair result even in agreements between two private parties. "It would be a great mischief to the community, and a reproach to the justice of the country," Chancellor Desaussure of South Carolina argued in 1817, "if contracts of very great inequality, obtained . . . by the skillful management of intelligent men, . . . could not be examined into, and set aside."

Gradually, the equitable theory of contract gave way to a "will" theory, which upheld private bargains made by individuals, regardless of their intrinsic justice. The single case of *Seymour v. Delancey,* decided differently by two courts, dramatically exemplifies this transition in legal doctrine. In 1822, Chancellor James Kent of New York refused to uphold a contract exchanging two farms for two town lots. "The village lots were not worth half the value of the country farms," he wrote, and such a "hard bargain" was not "fair and just." Two years later, a higher state court reversed Kent's verdict. This court argued that the parties had freely entered into the transaction and were in the best position to determine its fairness. "Purchases are constantly made upon speculation," the court pointed out, "[and] the value of real estates is fluctuating." Hence, it concluded, courts should intervene in private contracts only when there was "flagrant and palpable" fraud.

This innovation in legal theory, in combination with changes in the laws of property and negligence, facilitated the expansion of the market economy and the triumph of a capitalist system of business ethics. In *The Wealth of Nations* (1776), the Scottish economist Adam Smith posited the existence of an "invisible hand" that brought buyers and sellers previously unknown to one another together in large-scale markets and regulated their transactions. In fact,

Smith's "invisible hand" consisted of the legal rules of contracts and of the courts that enforced them. Therefore, the newly powerful judicial system created the institutional and ideological foundation of the emergent market economy.

The creation of the legal foundation of a capitalist society was a major achievement of the years between 1790 and 1820. With the national government, the systems of public finance, the political parties, and the policies of state mercantilism, it formed a new institutional matrix. This new and complex political framework would shape the identities and aspirations of millions of Americans in the materialistic and nationalistic world of the nineteenth century.

10

The Dynamics of Economic and Social Change, 1790–1820

In 1775, the character of American society was very similar to what it had been in 1700. In New England and the middle colonies, freeholding farm families still formed a majority of the population. Their localist outlook and egalitarian social values shaped the character of rural life. Continuity characterized southern society as well. Wealthy white planters continued to profit from the tobacco, rice, and wheat produced by enslaved blacks and, by controlling land and credit, dominated the lives of many white yeomen and tenant families. The size and complexity of these two preindustrial societies had increased, but their structure had not fundamentally changed.

Then came a period of far-reaching social and economic evolution. This transformation began slowly in 1775 and then, after 1790, moved rapidly ahead. By 1820, the United States had become a highly developed preindustrial society with a diverse economy and a complex social structure.

The rapidly shifting mosaic of American life contained four distinct pictures of economic and social change. First, the export-oriented seaboard economies of the Chesapeake region and the Carolinas remained important until 1820, and northern merchants became more prosperous than ever before. As a result, Virginia planters and Federalist merchants continued to be central actors in American social and political life.

Second, American settlers carried their traditional systems of economic production into the trans-Appalachian west. To supply the ever increasing demands of English textile mills for cotton, southern planters transplanted the slave labor system far to the south and west of its place of origin. Northern farmers carried their freeholding agricultural system and egalitarian social values into the fertile lands of the Ohio and Mississippi river valleys. By 1820, western freeholders were shipping thousands of bushels of wheat to markets in the eastern states and in Europe. This geographic expansion involved significant human costs. A new series of Indian wars nearly destroyed the native American tribes of the region, and the forced transplantation of slave society undermined the family and community life of thousands of black Americans.

A third set of economic changes affected many white Americans. The growth of household manufacturing altered the character of rural society in the northeastern part of the United States. Simultaneously, commercial expansion prompted the rapid growth of towns and cities. The increasing prominence of a market economy, for labor and domestically produced goods as well as exports, created a full-fledged capitalist system in many areas and a society that was sharply divided between wealthy owners and propertyless workers.

Finally, the dynamic economic changes, along with shifting cultural values, changed the nature of the social system, especially in urban areas. Residents of towns and cities formed new religious organizations, professional societies, and voluntary associations. Some Americans forged new types of family relationships and began to create a distinct middle-class culture. By 1820, American life was characterized by a new social complexity arising from geographical expansion, internal capitalist development, and an urban culture.

Surge and Decline in the Export Economy

By preindustrial standards, the American agricultural economies were an outstanding success. European travelers commented favorably on the prosperity enjoyed by the majority of Americans. After observing poverty in Europe, Thomas Jefferson agreed. "Of twenty millions of people" in France, he wrote from Paris in 1787, "there are nineteen million more wretched, more accursed in every circumstance of human existence than the most conspicuously wretched individual of the whole United States." Like most whites, Jefferson simply ignored the 750,000 Americans (nearly 20 percent of the total population) who were blacks or slaves. He was on surer ground with respect to the white population. The annual per capita income of whites in the British mainland colonies in 1774 was about $71, very close to the per capita income of $100 achieved in the more highly developed English economy.

The higher English income was a measure of the limitations of the American agriculturally based economies. By 1750, tens of thousands of English families manufactured textiles in rural households or made pottery and iron utensils in small factories. By 1811, only 35 percent of all English families were directly engaged in farming, compared with 84 percent of the American population.

This difference accounted for the higher English per capita income, since the yearly output of a nonagricultural worker was nearly double that of a farm laborer.

A strong American export trade in farm commodities compensated somewhat for the lack of domestic manufacturing. Rising prices for tobacco, wheat, and rice largely accounted for American prosperity between 1750 and 1775. The War of Independence brought a sharp decline in exports and in living standards, especially in the seaport cities, southern tobacco plantations, and northern areas of commercial farming. Prosperity returned only with the revival of international trade in the late 1780s. By 1793, the per capita value of American imports finally regained the level of the early 1770s. Then, during the Wars of the French Revolution (1792–1815), the value of imports shot dramatically upward.

The import boom helped to raise the American standard of living to a new high. Significantly, the United States paid for these imported textiles, iron goods, and other manufactured items out of current earnings rather than by incurring new debts. The American foreign debt amounted to $61 million in 1790 but only $62 million in 1805. Yet merchants imported goods valued at $700 million during the intervening period.

As in the past, agricultural exports paid for most of these manufactured imports. Increasing quantities of flour, tobacco, rice, meat, and, especially, cotton flowed out of American farms and plantations to markets in Europe and the West Indies. These farm products accounted for 80 percent of commodity exports, and the extractive industries of lumbering (11 percent) and fishing (5 percent) contributed most of the balance. Because of rising prices between 1788 and 1818, these traditional types of exports were more valuable than ever before. The total annual value of commodity exports rose from $20 million in the 1790s to nearly $70 million in the 1820s, as Table 10.1 on page 224 indicates.

The population of the United States increased almost as rapidly (from 3.9 to 9.6 million between 1790 and 1820) as the value of exports, so that exports alone could not support a higher standard of living. Between 1792 and 1815, the freight earnings of American merchant vessels provided additional national income. Continual warfare between France and Britain nearly destroyed the merchant fleets of both countries. Exploiting this wartime opportunity, American merchants took over the trade between Europe and the West Indies. To meet the technical requirements of international law, they carried sugar and other products from Caribbean plantations to American ports. Then, after paying and receiving a refund of United States customs duties, they "reexported" these goods to their European destinations.

Wartime profits were immense. American freight earnings averaged more than $20 million a year between 1793 and 1802, compared to $6.5 million a year between 1790 and 1792. And they rose to a spectacular annual average of $31 million for the ten years beginning in 1803. For more than two decades, such earnings from shipping constituted the largest share of American foreign

TABLE 10.1
United States Balance of Payments, 1790–1819
(in millions of dollars)

Year	Exports & Ship Sales	Imports & Specie	Freight Earnings	Insurance & Interest Costs	Yearly Balance	Existing Debt
1790	20.1	−22.5	5.9	−4.6	−1.1	−61.1
1800	22.6	−42.1	26.2	−8.5	−1.8	−82.3
1810	43.3	−66.9	39.5	−9.1	6.8	−72.9
1819	70.7	−92.5	15.2	−8.1	−14.7	−176.0*
1790–1819	1,192.1	−1712.8	628.8	−224.1	−116.0	

SOURCE: Douglass C. North, "The American Balance of Payments, 1790–1860," in *Trends in the American Economy,* Table A-4.

*Includes British commercial debts of $88.5 million incurred mostly after 1815 and repudiated during the Panic of 1819. The resulting balance of $87.5 million was comprised largely of the debts of the United States government.

Earnings from exports and shipping allowed Americans to import huge quantities of manufactured goods from Britain between 1790 and 1810 without falling deeply into debt. The subsequent decline in freight earnings brought a rapid rise in commercial indebtedness.

exchange — more valuable than any single farm export, including tobacco and cotton.

Commercial profits revitalized the life of the seaport cities. Merchants expanded their warehouses and built expensive new residences. Rows of three-story "Federal-style" mansions lined Chestnut and Essex streets in Salem, Massachusetts, and in similar residential districts in all the port cities of New England. This building boom gave work to hundreds of carpenters, masons, and furniture-makers (and explained their loyalty to the merchant-led Federalist Party). Thousands of artisans and laborers also found jobs in the rapidly growing ship-building industry. In fact, ship construction emerged as the premier American industry. Domestic shipyards produced most of the vessels that raised the tonnage owned by American merchants from 355,000 tons in 1790 to 1,089,000 tons in 1808. The increased availability of jobs was responsible for a permanent rise in the urban population. By 1810, over 10 percent of the inhabitants of New England and the Middle Atlantic states lived in towns of 2500 or more. This surge of output and urban growth testified to the vitality of the export-oriented economy of eighteenth-century America.

However, the weak foundations of this productive system soon became apparent. The commercial boom of 1790–1815, like that of the 1750s, stemmed in part from unusually favorable international conditions. In 1807, Jefferson's embargo cut the flow of American exports as well as the reexport trade in West Indian goods. Peace in Europe caused this temporary decline to become a permanent fact of life. Between 1815 and 1845, American merchants earned only $10–15 million per year for shipping freight — a mere third of the peak wartime level. Peace also brought the revival of European farming and a sharp drop in world agricultural prices. In the crucial year 1818, the market price of a pound of American cotton dropped from 33 cents to 20 cents, and that of a barrel of flour fell from $10 to $7.

This precipitous drop in farm prices sparked the first nationwide economic recession in American history, the so-called Panic of 1819. With falling incomes, farmers could not pay debts owed to merchants (for seeds, farm equipment, and clothes). Some merchants filed lawsuits to seize the assets of their farm customers. Many traders could not repay their British suppliers or their loans from American banks and therefore declared bankruptcy. Unable to collect these loans, many state-chartered banks closed their doors. In essence, the Panic of 1819 represented the collapse of the credit system of American commercial agriculture, a disaster triggered by rapidly falling world farm prices.

The Panic of 1819 marked the end of the predominance of the export-oriented economy. Southern planters and northern farmers continued to ship their products to foreign markets, in increasing absolute amounts. But the *relative* importance of foreign trade declined sharply. During the 1820s, the per capita value of imports dropped below the level of the 1770s. (See Chart 10.1 on page 226.) The United States subsequently would obtain a rising standard of living for its citizens through internal economic development and western geographic growth.

Western Growth and Native American Decline

Throughout European history, two traditional obstacles hindered economic growth: static productivity in manufacturing and "diminishing returns" in agriculture. Population growth was the immediate cause of diminishing returns. As the number of inhabitants in a village or region increased, some peasants had to cultivate upland fields with poorer soils or to plant their best fields every year, depleting the fertility of those fields. In either case, the work of each laborer yielded a smaller harvest.

The settlement of America represented an attempt to break out of this cycle of diminishing returns. For more than a century, it worked. By 1750, however, long-settled plantations in the West Indies and the southern mainland colonies were yielding less sugar and tobacco each decade. And heavily populated farms in New England and the middle colonies had to be worked more intensively to produce sufficient crops of wheat, corn, and potatoes. These lands were not biologically exhausted; they simply required greater inputs of capital or labor to obtain a given yield. Geographic expansion into the interior of North America offered a new opportunity to break out of the cycle of diminishing returns. In economic terms, the natural resources of the trans-Appalachian west were "free goods."

CHART 10.1
Decline in the American Export Economy, 1790–1830

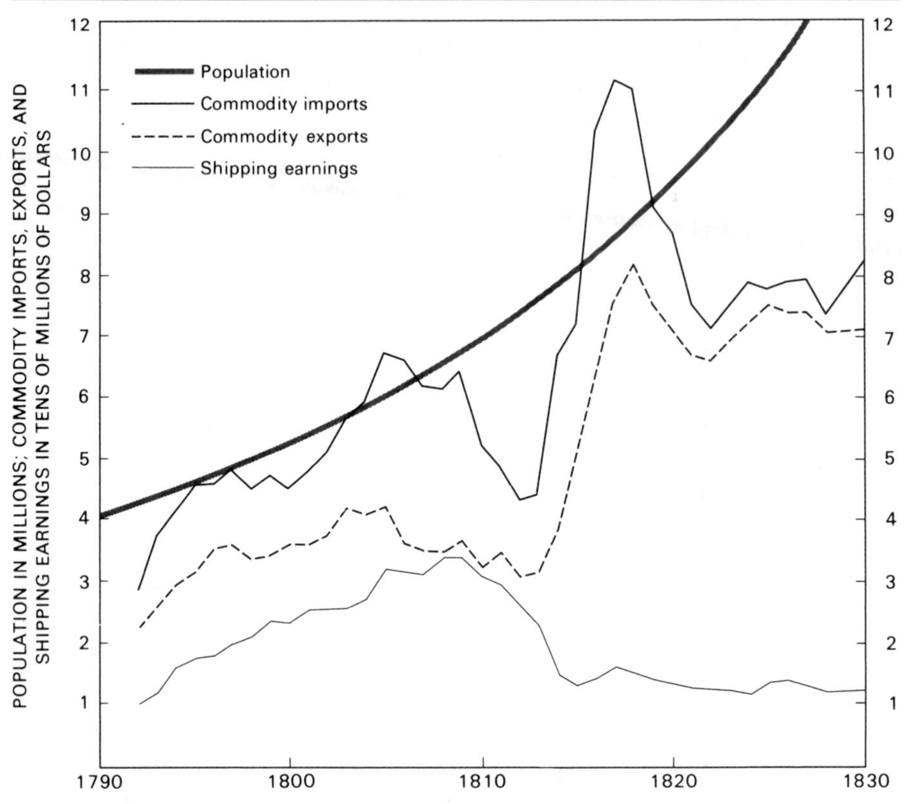

SOURCE: Douglass C. North, *The Economic Growth of the United States, 1790–1860* (Norton, 1966), Tables, A–III, B–III, E–III, B–VIII, C–VIII.

After 1820, neither foreign trade nor shipping earnings kept up with population growth. American prosperity began to depend increasingly on domestic production and commerce.

In reality, the exploitation of this land required the military subjugation of the Indians who lived there. Americans began moving into the trans-Appalachian west in the 1750s, sparking the French and Indian War and numerous local clashes with Indian tribes. Native Americans successfully resisted this invasion into their ancestral lands for a generation. Then, during the War of Independence, American military superiority shifted the balance of power. To protect their lands, the powerful Iroquois of New York allied themselves with the British, as did most of the tribes to the south and west. Hundreds of Indian warriors died, and scores of native American villages were devastated. By the end of the war, many tribes could no longer resist white expansion. Moreover, in the peace negotiations, the British did not protect the interests of their Indian allies, as they could have done by demanding the creation of an Indian "buffer state" south of Canada. In the Treaty of Paris of 1783, the British relinquished sovereignty over the entire trans-Appalachian west to the United States.

For their part, American political leaders treated the western Indians as conquered peoples. In the Treaty of Fort Stanwix (1784), congressional commissioners dictated the terms of settlement with the Iroquois, Delaware, and other tribes. Backed by a detachment of troops, these commissioners assigned specific geographic boundaries to each Indian tribe. Once out of the range of American guns, most of the tribes repudiated the Stanwix agreement. Many of these tribes soon split into bitterly divided factions. Some native Americans refused to leave their ancestral villages on any terms. Other tribal members negotiated new treaties that relinquished portions of their domain in return for money and manufactured goods. Between 1798 and 1806, for example, one faction of Cherokee chiefs ceded lands to the United States in four separate treaties, thereby angering other members of the tribe. An opposition faction took revenge in 1807; accusing Chief Doublehead of plotting another cession of Cherokee lands, they killed him.

In New York, national and state officials exploited similar divisions among the Iroquois. They showered goods and money on the Iroquois leaders who would sell parts of their homeland; then they assisted white settlers to drive out those clans that refused to move. Full-blooded Indians led by the Seneca chief Red Jacket opposed land cessions, but without much success. Confined on smaller and smaller reservations, the once vigorous Iroquois declined into alcoholism and despair. Then, between 1799 and 1815, Handsome Lake, a charismatic prophet, led a religious revival. Combining traditional Indian teachings with Christian principles, Handsome Lake devised a new religion that sustained the moral strength and social identity of the Iroquois people.

The undefeated tribes of Ohio and Indiana were less easy to subdue. Organized in the Western Confederacy, the tribes defeated United States troops in 1790 and 1791. Bolstered by substantial reinforcements, General Anthony Wayne defeated the confederacy in 1794 in the Battle of Fallen Timbers. Even so, the Treaty of Greenville (1795) only prohibited the Indians from alienating their lands to a foreign nation. Until the tribes "shall be disposed to sell," the treaty guaranteed them "the quiet enjoyment of their lands, against all citizens of the United States."

In fact, this agreement only postponed the day of reckoning. White settlers flooded into the trans-Appalachian west, and government officials forced cessions from the tribes. As governor of the Indiana Territory, General William Henry Harrison used bribes and favors to purchase millions of acres of tribal lands. In the Battle of Tippecanoe in 1811, Harrison defeated the great Indian leader Tecumseh. This Shawnee chief had resurrected the Western Confederacy in a last-ditch effort to halt white settlement at the Ohio River.

In the South, General Andrew Jackson defeated the Creeks in 1814 in the Battle of Horseshoe Bend. He forced the tribe to cede 20 million acres in Alabama and Georgia. Jackson extracted millions of additional acres in Tennessee

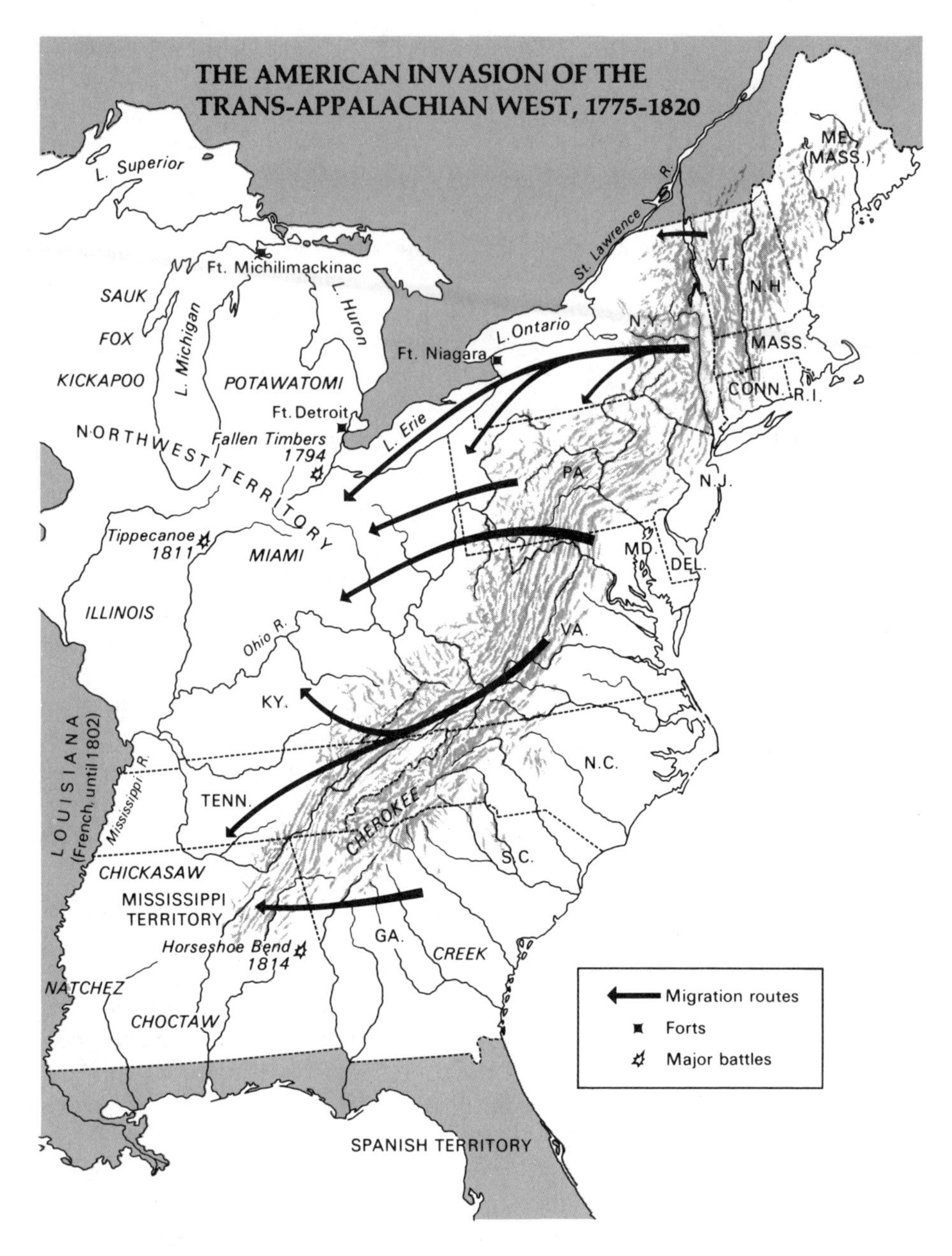

Between 1790 and 1820, nearly two million Americans settled in the trans-Appalachian West, sparking a new series of Indian wars. In the following decades, the United States government used legal means and military force to move most native Americans to "Indian Territories" west of the Mississippi River.

and Mississippi from the Choctaw and Chickasaw. Then Jackson turned against the Cherokee, his military allies at Horseshoe Bend, forcing them to sell 3 million acres. To resist white expansion, the Cherokee created a strong tribal government, adopted a constitution in 1827, and defended their legal rights before the United States Supreme Court. In spite of these efforts, the Indian Removal Bill of 1830 sealed the fate of the Cherokee and other southern tribes. By 1840, United States military forces had driven most Indians to reservations west of the Mississippi River.

Southern planters accompanied by thousands of black slaves quickly moved onto native American lands in Georgia, Mississippi, and Alabama. By exploiting these rich soils, planters not only averted diminishing returns but also propelled the southern plantation economy to new heights of prosperity. Annual cotton production rose from 3 million pounds in 1793 to 93 million pounds in 1815. From 1815 until 1833, cotton comprised one-third of the value of all American exports, and it accounted for one-half of all exports in the three decades before the Civil War. Besides exporting cotton, southern planters supplied it to the textile mills of New England and the Middle Atlantic states, greatly assisting the economic development of those regions.

The settlement of western lands promoted economic development in the seaboard states. Nearly a million people had migrated into the western counties of New York by 1820, and 400,000 more were already settled in the new states of Ohio (1803), Indiana (1816), and Illinois (1818). Initially these migrant farm families raised grain and meat, which they sold to newly arrived settlers. In return, they received cloth, farm equipment, shoes, and other manufactured goods carried from the eastern states. By the 1820s, however, western farmers had become direct participants in the national market. Canals and turnpikes recently built by the state governments provided cheap, reliable transportation. Using the Erie Canal, farmers in Buffalo and Rochester shipped thousands of barrels of flour a month to New York City. Barges carried grain and pork down the Ohio and Mississippi rivers for shipment to Europe or eastern cities, and newly built fleets of steamboats transported manufactured goods back up those rivers.

The appropriation of Indian lands and state mercantilist policies combined to allow western farmers and southern cotton planters to escape the trap of diminishing returns.

Internal Development and Household Production

Even as many Americans migrated westward, the residents of the northern seaboard states were struggling to get past the second traditional obstacle to economic prosperity, static productivity in manufacturing. Americans began this effort during the War of Independence. To ensure adequate supplies for its military forces, the Patriot movement in Connecticut and elsewhere urged merchants to promote the domestic manufacture of shoes, gunpowder, and cloth. Four years later, a French military officer, the Marquis de Chastellux, visited a farmstead in Farmington, Connecticut. "The sons and grandsons of the family were at work," he reported, "making a sort of camblet, as well as another woolen stuff with blue and white stripes for women's dress. . . . One workman can make five yards a day." Encouraged by the state government of Maryland, a clothing contractor set up a small factory with sixteen looms; soon it was producing 100 yards of linen per day.

The war increased domestic output of cloth and changed the character of production. Previously, women and girls had spun wool and flax into yarn, which male weavers made into cloth. During the war, men were called into military service, and the demand for cloth rose, prompting many women to take up weaving. "Among the poor the wife weaves generally," Thomas Jefferson noted, "and the rich either have a weaver among their servants [i.e., slaves] or employ their poorer neighbors." As women expanded their productive roles, male artisans diversified their activities. They fabricated guns, gunpowder, paper, iron utensils, and shoes. Because a greater number of people, female as well as male, were being employed in the more productive manufacturing sector of the economy, per capita output grew. As David Ramsey of South Carolina noted, the war had encouraged — at times, even forced — Americans to expand their working lives "in a line far beyond that to which they had been accustomed."

The war had a lasting effect on American economic development. Immediately following independence, American merchants once again began to import huge quantities of British goods. The influx of cheap, well-made imports cut the profits of new domestic textile factories; both the Hartford Manufactory (founded in 1788) and the Massachusetts Woolen Manufactory (1794) soon fell victim to British competition. However, American household production expanded, as artisans and farm families produced an increasing diversity of manufactured goods. In the small village of Hallowell, Maine, during the 1780s, for example, the young daughters of Martha Ballard learned to weave a wide variety of fabrics. Other village women wove, too, so that by the 1790s, half of the probate inventories in Hallowell listed looms among the family property. Weaving became a normal household responsibility for women in Ulster County, New York, as well. Cloth production in Ulster climbed to 6.1 yards per capita in 1820 and reached 8.6 yards in 1825; by that time, the county's household producers provided work for thirty mills that "fulled," or shaped, the woven woolen cloth.

Most households produced cloth for their own use rather than for sale. Nevertheless, this widespread domestic textile manufacturing had profound economic consequences. It encouraged the keeping of sheep (for meat as well as for wool), prompted the building of fulling and spinning mills, and diminished the need for British imports. The British consul at Philadelphia reported with concern in 1789 that "among the country people in Massachusetts coarse linens of their own making are in such general use as to lessen the importation of checks and even of coarse Irish linens nearly ⅔rds."

The expansion of household manufacturing between 1775 and 1820 laid the foundation for a self-sufficient and prosperous America. During a visit to Berlin, New Hampshire, in 1797, traveler Timothy Dwight learned that before independence locally produced tinware had been sold by peddlers with a horse and two baskets. He was told:

> After the war, carts and wagons were used for this purpose. A young man is furnished by a proprietor with a horse, and a cart covered with a box. . . . This vehicle within a few years has, indeed, been frequently exchanged for a wagon; and then . . . these young men direct themselves to the Southern States.

New England shoe manufacturers were another group who increased their output for market sale. During the war, artisans in the small town of Lynn, Massachusetts raised their annual production of shoes to supply American military forces and southern slave plantations (which could not get shoes from Britain). By 1789, the town turned out 175,000 pairs of shoes each year, and by 1800, no fewer than 400,000 pairs. With these and other increases in domestic manufacturing, the United States transcended the export-oriented economy of the eighteenth century. The nation became able to produce many manufactured goods that previously had to be imported from Great Britain.

Three factors played key roles in the expansion of household manufacturing. First, American women and their husbands were choosing to have fewer children. Wives gave birth to four or five children on the average, whereas their grandmothers had commonly had seven or eight. With fewer pregnancies, fewer babies to nurse, and fewer children to clothe and feed, women had more time and energy to devote to producing goods — both for their families and for market sale. "Along the whole road from Boston," a visiting Polish aristocrat noted in 1797, "we saw women engaged in making cheese."

Year-round employment constituted a second major cause of the increased productivity. Before 1750, and even up to 1775, the seasonal cycle of the agricultural year determined the rhythm of economic life. Men and women worked hard only during peak periods of farm activity. During mid-summer weeks and winter months, they had little to do. A growing number of "by-employments," or off-season labor, increased the productivity and the incomes of many farm families. A traveler in Massachusetts noted:

> In the winter season, the inhabitants of Middleborough are principally employed in making nails, of which they send large quantities to market. This business is a profitable addition to their husbandry; and fills up a part of the year, in which, otherwise, many of them would find little employment.

Energetic entrepreneurial merchants and artisans were the third ingredient in the expansion of household manufacturing. During the 1780s, country traders actively sought goods from farm families and carried them to market on the newly built turnpikes. Simultaneously, town merchants and master artisans developed networks of rural "outworkers." They employed farm women and children to sew the soft uppers of shoes, to spin yarn, or to make straw hats.

Ebenezer Breed of Lynn, Massachusetts, was a typical entrepreneur. In 1792, he formed a manufacturing partnership with a fellow shopkeeper. Breed traveled constantly, purchasing leather hides and finding customers. He also petitioned the United States government for greater tariff protection for domestic shoemaking. Meanwhile, his partner in Lynn supervised the production process. He hired local artisans to cut the leather, wagon drivers to carry it to the backcountry, rural outworkers to make the uppers, and other local shoemakers to attach the soles. By developing such a complex putting-out and marketing system, Breed (and other entrepreneurs) raised the annual output of shoes in Lynn to nearly a million pairs by the 1820s.

The character of the eastern agrarian economy had changed. By 1800, local traders and urban merchants had created a dynamic *intra*regional market in farm produce and household-manufactured goods. Consequently, many eastern communities were well prepared to take advantage of the new opportunities offered by *inter*regional trade. As cheap western farm commodities flooded their markets, some eastern farmers specialized in the production of more perishable goods: milk, cheese, and fruits. Other farmers raised sheep or cattle, profiting from the demand for wool and leather from textile and shoe factories. Eastern agriculture became more efficient, producing more specialized goods with fewer laborers.

Not all eastern farm families participated in this more competitive and market-oriented agricultural economy. Some deserted their rural villages for urban life; others sought to perpetuate their traditional way of life on new western farms. Many eastern rural folk found a compromise solution that provided economic security without disrupting the ties of family and neighborhood. In many long-settled regions, such as Lancaster County and Chester County in Pennsylvania, 30 to 40 percent of the rural residents on the tax rolls were laborers or artisans. They lived on the land and grew their own crops; but they, or their wives and children, derived additional income from outwork or the practice of a trade. As Alexander Hamilton noted proudly in 1792, the countryside was "a vast scene of household manufacturing . . . in many cases to an extent not only sufficient for the supply of the families in which they are made, but for sale, and even, in some cases, for exportation."

The household system of manufacturing increased productivity without creating a depressed class of propertyless workers. The family remained the primary economic unit. Men and women worked in their own homes or in those of neighbors. They set the pace of their labor, avoiding the harsh discipline of factory work. Equally important, many of these farm families avoided overdependence on the outwork system. Living on small subdivided farms, they took advantage of the opportunities provided by merchant capitalists to maintain a sufficient income, but did not rely on outwork for basic subsistence. Like the yeomen farm families they came from or still lived near, these household manufacturers wanted to live in republican communities composed of relatively prosperous and independent property owners. Between 1790 and 1820, many families and communities in New England and the Middle Atlantic states achieved this goal. They expanded their output of farm goods, timber, and manufactured items and increased the volume of trade among themselves. This process of internal economic development preserved and even made more prosperous the traditional society of yeomen farmers.

Toward a Capitalist Society

The year 1820 represented the high-water mark of this republican society of small-scale household producers. Already the expansion of the capitalist outwork system and the advent of factories had begun to undermine the social world of these communities of economically independent men and women. In 1790, Samuel Slater, an English immigrant, formed a partnership with Moses Brown, a wealthy Rhode Island merchant and entrepreneur. Utilizing English technology, Slater set up a cotton manufacturing company. Slater, Brown, and their capitalist associates owned the factory, the machinery, and the raw materials. They hired entire families to tend the machines that cleaned and combed

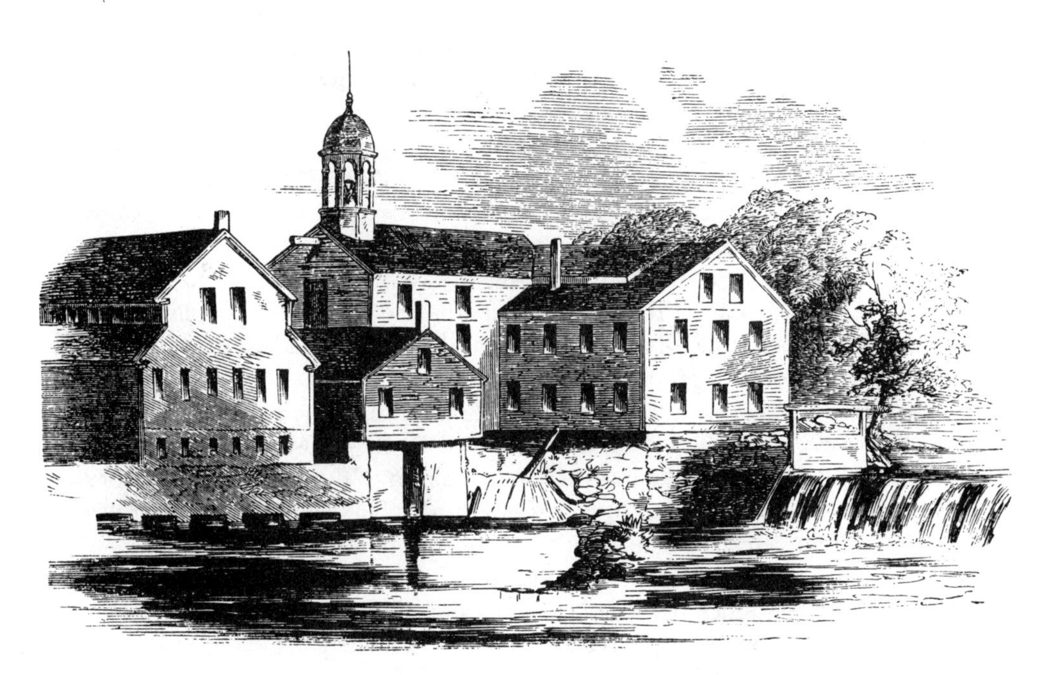

Slater Mill in Pawtucket, Rhode Island. This early textile factory used waterpower to drive spinning machines that had dozens of bobbins. Productivity increased dramatically, but most textile workers did not receive any significant share of the profits generated by the new technology. (*Source: The Rhode Island Historical Society.*)

the cotton and then spun it into thread. These workers were no longer independent producers but propertyless wage-earning proletarians.

The system of factory production grew steadily. Initially, most textile factories simply spun the wool or cotton into thread, sending it to household workers to weave into cloth. In 1809, six textile factories in Philadelphia produced 65,000 yards of material, while nearby household manufacturers achieved an output of 230,000 yards. By 1812, there were eighty-seven cotton mills in the United States, and many of them contained looms as well as water-driven spinning machines. They employed 500 men and 3500 women and children. The expansion of factory production deeply affected the character of nineteenth-century life. Each year, more families left the countryside to work in new mill towns or commercial centers. As propertyless wage-earners, they had next to no control over the conditions of their work. They worked long and hard, tending machines that ran from dawn to dusk for 6 days a week, 309 days a year. In addition to Sundays, New England textile workers had only three regular holidays: Fast Day (a Puritan holy day) in the spring, the Fourth of July, and Thanksgiving.

At first, rural migrants did not readily submit to the monotonous and severe discipline of the factory. Like their yeomen forebears, they preferred freedom and leisure to greater earnings. When Lucy Larcom quit her job in a Massachusetts textile mill, the paymaster inquired if she was "going where you can earn more money." "No," Larcom replied, "going to where I can have more time." As a young single woman, Larcom had more options than married workers with small children. To provide for their families, parents had to accept whatever work was available, even if it drained their physical strength and crushed their spirits. In America as in England, as Edward Thompson has noted, it was "neither poverty nor disease but work itself which casts the blackest shadow over the years of the Industrial Revolution."

The new American working class might have accepted the factory system with greater enthusiasm if the financial benefits had been more equally distributed. Although some factory workers achieved a higher standard of living, capitalist entrepreneurs appropriated the bulk of the profits. In Philadelphia in 1774, the merchants who made up the richest tenth of the population owned about 50 percent of the city's taxable property. By the mid-nineteenth century, wealthy capitalists enjoyed a much greater position of dominance; a small proportion of families — the top 1 percent of Philadelphia's taxpayers — controlled no less than 54 percent of the wealth.

In the new urban industrial world, the lines of class division were much more definite and much harder to cross. In a detailed survey of Philadelphia in 1830, Reverend Joseph Tuckerman reported that

> . . .the classes . . . who are wholly dependent upon wages are very numerous . . . and to a great extent are living, as a caste, — cut off from those in more favoured circumstances, and doomed to find their pleasures, and sympathy in their suffering, alone among themselves.

Moreover, property-owning artisans no longer formed a significant segment of urban society. Even before the coming of the factory system, market pressures undermined the viability of many crafts. Merchants bought shoes, clothes, farm implements, and iron goods produced in factories or in other regions of the country, so that local artisans no longer enjoyed a monopoly. To survive, they had to produce goods at a competitive price.

To combat the impersonal forces of the market, craft workers pursued two policies. First, they attempted to exclude low-priced foreign imports. On July 4, 1788, calico-printers joined other Philadelphia artisans in the parade celebrating the ratification of the Constitution. Their intricately patterned flag carried the motto "May the union government protect the manufacturers of America." Second, journeymen artisans tried to impose uniform wage rates on the master craftsmen or merchants for whom they worked. In 1805, Philadelphia shoemakers demanded wage parity with New York cordwainers. Journeymen in the printing industry subsequently established uniform standards and pay scales to prevent cutthroat competition.

The struggle to moderate the impact of the market economy continued throughout the nineteenth century. Some artisans secured tariff protection for their handmade goods; others extracted uniform wage rates from their employers. In general, however, artisans could not compete against factories. Machines tended by unskilled or semiskilled workers could produce more goods — and often better goods — more cheaply. Thus, the mechanization of production destroyed many traditional artisan crafts. At the same time, the factory regime created a demand for other types of skilled craftsmen — tool makers and repairers, construction workers, and millwrights. On the whole, however, the proportion of self-employed, property-owning artisans declined over the course of the nineteenth century. A distinct social group was gradually squeezed practically out of existence.

The loss of the artisan class was notable because artisans stood for a distinct set of social values. Along with yeomen farmers and rural household manufacturers, artisans saw themselves as the "producers" of society — the men and women who used their hands and minds to raise crops and to fashion goods. Like Thomas Jefferson, most artisans advocated a republican social order composed of independent small-scale producers. They feared the division of society into two classes, wealthy capitalists and wage-earning proletarians. And many questioned the capitalist ethic of unrestrained competition and unlimited accumulation. One Philadelphia journeyman explained:

> A conscientious man in these times, can scarcely expect to earn more than a *competency;* if more than this is aimed for, man is apt to become the oppressor of his fellows — taking advantage of their necessities and obtaining the fruits of their labor without rendering a just recompense.

The factory system undermined the artisan economy and its producer ethic. In the new capitalist system of economic organization, there were proportionately fewer independent entrepreneurs who controlled productive property. Ordinary workers could aspire to rise up through the ranks within a manufacturing concern, winning a foreman's position through devotion and hard work. Supervisors and clerical workers could similarly look forward to the day when they could use their savings and experience to establish a small enterprise of their own. Men and women continued to hope and dream; their optimism, however, could not eliminate the obstacles to occupational mobility created by the factory regime.

The social cost of industrialization was offset (to the extent that fundamental changes in human opportunities and values can be reduced to the terms of a balance sheet) by the appearance of a more productive economic system. Capitalist-controlled markets and factories increased output and raised the material standard of living for many Americans. The alteration in the locus and character of manufacturing marked the appearance of a new social order. Eventually, the host of little producers, *families,* were replaced at the center of American social and economic life by a few massive groups, *classes* of employers and workers.

Regional Diversity and Social Complexity

These three dynamic processes of economic change — export expansion, western growth, and internal development — helped to create a more complex social order. By 1820, the United States contained six distinct regions, each with a different mix of social classes and cultural values. Three of these regions — the Northwest, Southeast, and Middle Atlantic — had many similarities to cultures that existed in 1775. The other three societies of the Chesapeake, Yankee New England, and the urban North were the products of a half-century of rapid social change.

Somewhat paradoxically, two of the more traditional societies existed on the American frontier. Contrary to the widely known interpretation advanced by the historian Frederick Jackson Turner in *The Significance of the Frontier in American History* (1893), the westward movement did not always produce new and more democratic, or more "American," societies. Rather, migration into the interior often allowed settlers to recreate their traditional regional social forms. When 176 members of the Congregational Church of Granville, Massachusetts moved to Ohio in 1815, they took their minister and their system of freehold agriculture with them. The migrants even selected a geographic site whose "peculiar blending of hill and valley" resembled the landscape of their former hometown. Throughout the Northwest — in upstate New York, Ohio, northern Indiana, and Illinois — settlers from overcrowded New England created farming communities that greatly resembled those of their pre-Revolutionary ancestors.

Southern migrants similarly replicated their traditional seaboard society, but in two distinct forms. The white farm families that streamed into the up-

REGIONAL DIVERSITY IN THE UNITED STATES, 1820

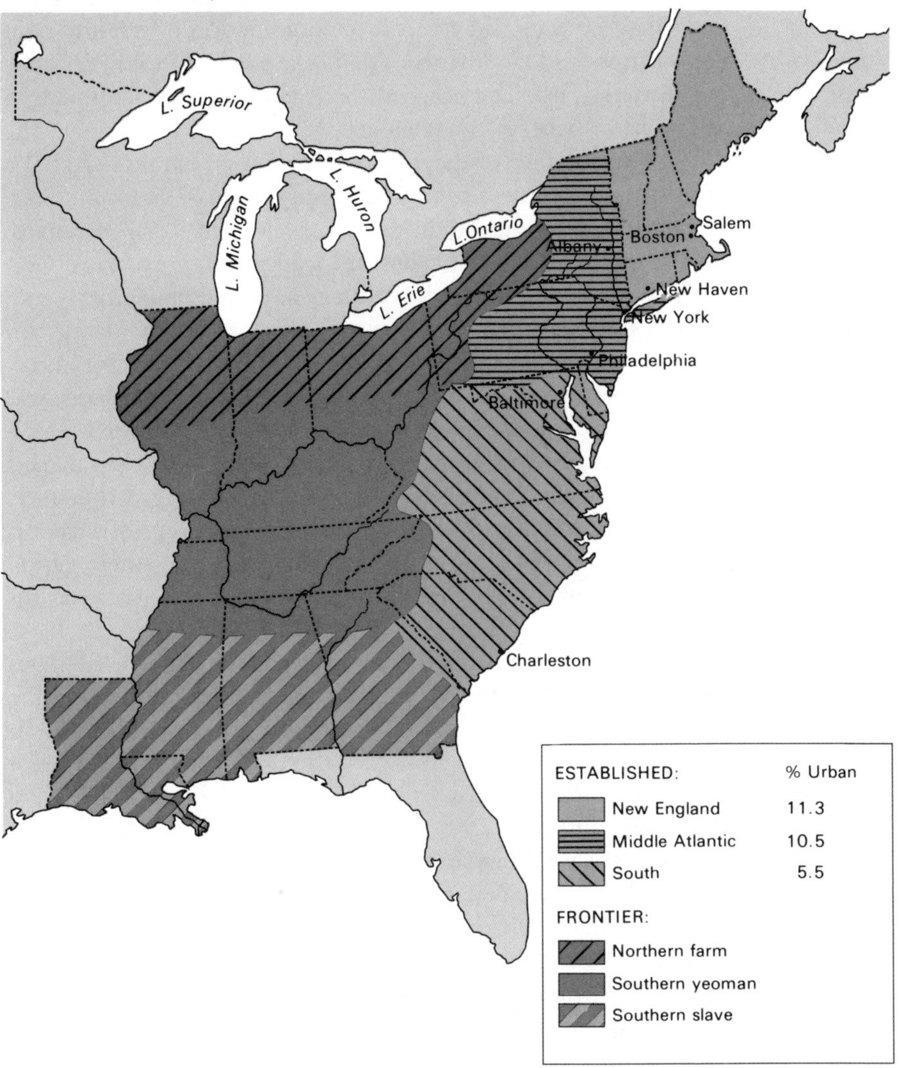

America remained a society composed of distinct regional cultures throughout the nineteenth century. New cultural values developed in seaboard areas, and migrants carried traditional regional housing styles, religious institutions, and cultural values into the trans-Appalachian West.

lands of Kentucky and Tennessee grew cotton for export, just as their ancestors had raised tobacco for foreign sale. But these migrants had significantly more economic independence, for they had escaped the domination of wealthy tidewater slaveowning planters and landowners. Many slaveowners followed a different migratory path westward. They carried the traditional slave regime into the flat and fertile "cotton belt" in Alabama and Mississippi. There, as in the Chesapeake and Carolina colonies of the 1740s, a few white families created a new patriarchal culture, growing wealthy on the labor of a mass of black slaves.

The third traditional culture in 1820, that of the Middle Atlantic region, still bore traces of the disparate ethnic origins of its inhabitants. Quakers, Germans, and Scotch-Irish continued to marry within their own ethnic groups and, especially in small agricultural settlements, maintained their cultural identity. In Hanover, Maryland, "the inhabitants are practically all German," a visitor noted, "habits, speech, newspapers, cooking — all German." Yet ethnic insularity was on the decline. Since 1775, few Europeans had migrated to the United States, so that most white Americans were now native-born. Moreover, the market economy and republican political and legal institutions encouraged the emergence of a common American cultural identity. Such an American identity was clearly developing; in 1820 the minister of the German Lutheran congregation in Frederick, Maryland, began to deliver his sermons and to keep church records in English.

Meanwhile, the society and culture in three other parts of the United States had become more complex and very different from what they had been in the eighteenth century. In the Chesapeake states, the fourth region that existed in 1820, the old tobacco economy had given way to the diversified production of corn, wheat, and livestock. Moreover, tens of thousands of former slaves who had been manumitted during the 1790s now lived with their children as free blacks in the states of Delaware, Maryland, and Virginia. Many of them, as well as many enslaved Afro-Americans, had developed strong family and community ties and belonged to Christian churches. Finally, their poor white neighbors also took their religious, and social, identity from membership in evangelical Methodist and Baptist churches, distancing themselves from the aristocratic culture of the gentry class. The old white planter elite continued to dominate the economic and political life of this region, but the lower orders had created their own vital cultures.

The society of New England had changed as well. The sixth- and seventh-generation descendants of the original Puritan settlers still dominated the communities of the region. For example, there were 259 members of the Breed family listed as residents of Lynn, Massachusetts, in 1830. However, many of them (like the shoe entrepreneur Ebenezer Breed) worked as manufacturers, either as household producers or as factory laborers. Puritan New England had become a land of "Yankees": sharp commercial bargainers, skilled artisans, and well-educated rural farmers and workers.

A new social system was even more apparent in the sixth region, the urban society of the United States. This diverse and class-divided culture was most prevalent in the New England and Middle Atlantic states. By 1820, over 11 percent of the population of these regions lived in towns of 2500 or more, and the large cities of Boston, New York, and Philadelphia had emerged as dominant economic and cultural centers. These urban centers were sharply stratified along the lines of wealth and class identity, but they also contained a wide variety of new professional, religious, and social institutions. It was within such growing towns and cities that the new American culture of the nineteenth century would take form.

The Context of Social and Religious Action

Before 1775, the lives of most Americans had been deeply rooted in the soil. Their allegiances fell into a natural hierarchy. In political terms, these men and women were primarily citizens of their village or settlement. They felt a weaker tie to the provincial government, and still less attachment to the British empire. Most Americans knew only a few individuals — merchants or former neighbors — in other settlements. Like the force of gravity, the circumstances of daily life pulled people's minds back to familial bonds, traditional religious ties, and the tiny sphere of life of the local village community.

In this society, the three "vertical" institutions of family, church, and community formed the prime contexts of social action. Each entity embodied similar patterns of authority. Husbands and fathers stood at the apex of the "little commonwealth" of family; by law and custom, they controlled the property and behavior of their dependents, their wives, children, servants, and slaves. Male heads of households elected some of their own number as their religious and political leaders. Once selected, ministers often presumed to speak with the authority of God, and gentlemen politicians demanded deference from the "lower orders."

As Old Lights and Loyalists complained, first the Great Awakening and then the American Revolution "turned the world upside down." New Lights and Patriots demanded religious liberty and political sovereignty for the people, and they were successful. The constitutions of the new American republics enshrined the twin doctrines of freedom of religion and sovereignty of the people. In addition, the struggle for independence created a sense of American political identity. Subsequently, state mercantilism increased the importance of state governments at the expense of those of towns, villages, and counties. By 1820, many white Americans were active participants in statewide or nationwide republican political systems and in an expanding market economy.

The context of religious and social action also changed. The old "vertical" institutions of family, church, and community did not vanish between 1775 and 1820, but they became less pervasive and authoritarian. For example, the American Revolution dramatically changed the traditional relationship between churches and the political state. Throughout the southern states, the Church of

England lost its privileged legal status. To win popular support for the war, the Virginia legislature passed James Madison's Act of Religious Toleration (1776), which extended legal rights to Presbyterian and Baptist churches. Ten years later, Thomas Jefferson's Bill for Establishing Religious Freedom made all churches equal before the law and made church membership a voluntary act. In the Middle Atlantic states, legislators similarly acknowledged the social fact of religious diversity by eliminating established churches and compulsory religious taxes. Even in New England, where Congregational churches maintained a privileged legal position until the 1830s, state governments allowed Baptists and Methodists to use religious taxes to support their own ministers. By 1790, the movement toward religious liberty and voluntary church membership that began with the Great Awakening of the 1740s had reached a successful conclusion.

In this free and open environment, churches emerged as prime examples of the voluntary association, a new type of social institution that undermined the traditional structure of authority. Church membership was determined by voluntary consent, and even small towns had two or three churches. Within each community, the compositions of the congregations often paralleled the levels of social class. Baptist and Methodist preachers appealed to poor tenant farmers, factory laborers, and craft workers. Often from humble backgrounds themselves, these evangelical ministers stood apart from their status-conscious colleagues. Yet increasingly, Baptist and Methodist clergymen commanded the religious allegiance of a majority of churchgoers. Ministers were no longer drawn from the top social stratum, and congregations no longer consisted of a cross section of the community.

The institutional reforms of the revolutionary era and the fluid and more diverse social world of the early nineteenth century had destroyed the "vertical" religious organization of one established church. In its place stood a host of "horizontal" organizations, voluntary associations of people with similar theological views or economic means. Presbyterian, Congregationalist, and Methodist competed with one another for converts and social influence. The values and practices of the market began to shape spiritual as well as economic life.

A similar trend toward occupational and professional diversity occurred within the ranks of the American social elite. In fact, upper-class groups had been the first to create horizontal institutions. The clergy had always been well organized, in voluntary Congregational "consociations" or in more hierarchical Presbyterian synods or in Anglican conventicles. Beginning in the 1740s, lawyers formed bar associations to define professional standards and to enhance their collective authority. Businessmen were equally active. Boston traders formed a merchants' club in 1763, and chambers of commerce appeared in many cities after the war.

Doctors created a professional identity with greater difficulty. As early as 1765, Dr. John Morgan of Philadelphia urged the establishment of a college of physicians. The college would control entrance into the medical profession by licensing doctors to practice in all the colonies. Morgan's scheme was too grandiose. Professional identity was too weak and American society too locally centered for a continent-wide organization to become established. Even the Medical Society of New Jersey, which received legislative approval to set standards for medical practice in 1772, had little success in doing so. On the eve of independence, nearly 3500 physicians practiced the "medical Art" in the continental colonies, but fewer than 400 had any formal training and only 200 actually held degrees.

By 1820, however, medical societies existed in every state. In many localities, they held the legal right to license doctors. The societies were especially active and effective in larger cities, where they set professional standards and established fee schedules. Public confidence testified to the new prestige and organizational strength of the profession. When yellow fever epidemics struck the seaport cities in the 1790s, residents turned to doctors for innoculations as well as to clergymen for spiritual aid.

A shift in social authority was occurring. The clergy no longer enjoyed an organizational monopoly. Other upper-class groups had developed distinct professional and occupational identities and began to compete with the clergy for social leadership. "Lawyers, physicians, professors and merchants were classes," the great historian Henry Adams proclaimed in his brilliant analysis of American society in 1800, "and acted not as individuals but as though they were clergymen and each profession was a church." Thus, the elite professions and the common people were no longer parts of a single, integrated, vertical social hierarchy but rather were members of distinct and often contending voluntary horizontal organizations. The character and the context of social and religious action had changed.

A Middle-Class Culture

By 1820, the sphere of American life had grown larger and more complex. The centrifugal forces of war and trade, geographic migration, and political and religious participation had shattered the tiny self-contained cosmos of the agricultural village. Farmers and artisans — as well as merchants, planters, and ministers — were more conscious of the larger social world in which they lived. They realized, if only dimly, that they depended on regional or international markets for economic prosperity, on the state or national government for tax and tariff legislation, and on national denominations for a sense of religious identity. Their perception shaped a responsive action. To exert control over this larger and more diverse social realm, Americans created an increasing array of voluntary institutions. The United States was the only country in the world, Alexis de Tocqueville announced in his famous study of American society, "where the continual exercise of the right of association has been introduced into civil life."

Voluntary associations reflected new republican values and a new middle-class pattern of social relations. Membership in these organizations was not hereditary or compulsory; individuals chose to join. Moreover, voluntary associations were not hierarchical, but were composed of "equals." Members usually elected their leaders. Sometimes the group selected those with inherited wealth or high status, but just as often they chose colleagues with political skills or forceful personalities. This emphasis on equality and on achieved (rather than inherited) status were important characteristics of a distinct middle-class culture.

In some respects, voluntary associations were the social analogues of republican political institutions, for their members retained ultimate authority. Unlike governments, however, voluntary associations did not include all members of a given community. Instead, they represented the interests or values of specific social groups. The Massachusetts Historical Society, founded in 1794, initially had thirty members. In addition to having a common intellectual pursuit, the society's founders were all men of wealth and influence. In fact, until 1800, most voluntary organizations had an upper-class membership. "The great easily form associations," a New Yorker noted in 1788, "the poor and middling classes form them with difficulty."

The lower orders had already begun to organize to some extent, however. Virginia tenant farmers flocked into Baptist churches during the 1760s, where they found social cohesion as well as spiritual sustenance. By joining Patriot organizations such as the Sons of Liberty and the Philadelphia militia, urban laborers and journeymen likewise created a collective identity. Similarly, master craftsmen formed patriotic associations and mechanics' clubs. During the 1790s, artisans dominated the new Democratic-Republican clubs and took an active role in the conflicts of the first party system. In Boston, newly-freed blacks formed the African Masonic Lodge, and free mulattoes in Charleston, South Carolina, created the Brown Fellowship Society. Led by Richard Allen, blacks in Philadelphia and other northern cities split off into their own Protestant denomination, the African Methodist Episcopal (AME) Church.

Previously, communities had exhibited a single hierarchy of status. Some families traditionally ranked at the top (or bottom) of the tax list; their assigned pews in churches indicated their rank in the social order. Private voluntary associations now formed a separate social hierarchy. Mechanics, merchants, lawyers, and bricklayers lived in the same community, but they increasingly inhabited distinct social worlds. Like the members of Boston's African Masonic Lodge, they associated primarily with others from their own occupation, class, or caste. The existence of this graded spectrum of social organizations was yet another characteristic of the new cultural system developing in urban areas of the United States. Social identity depended both on family background *and* on membership in a variety of associations. Unity had given way to multiplicity.

Religious participation and a heightened moral awareness also helped to shape the contours of the emergent middle-class culture. Beginning in the 1790s, a generation-long series of spiritual revivals transformed the character of American society. First, Universalists and Free Will Baptists brought together backcountry farmers in New England into new churches. Then, camp-meeting

CHART 10.2
Voluntary Associations: The Growth of Churches

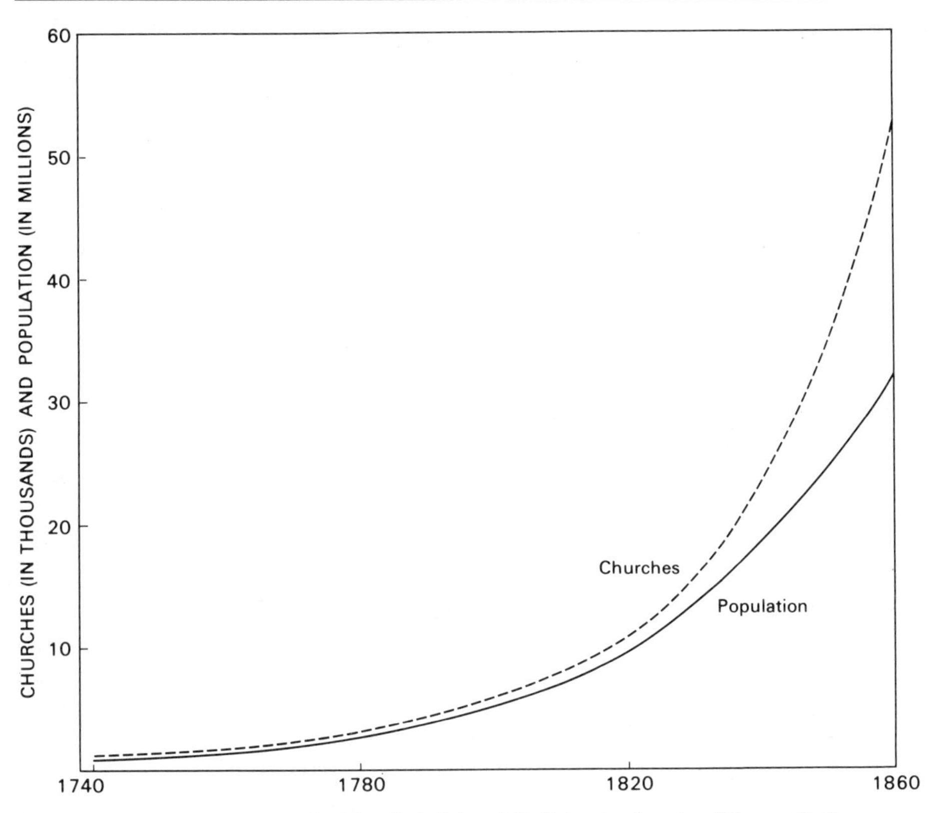

SOURCE: Edwin Scott Gaustad, *Historical Atlas of Religion in America* (Harper & Row, 1976), Figures 6, 31, and 32.

By 1820, the rate at which new church congregations were being established outstripped the rate of population growth. Most new church members became Baptists or Methodists, the most evangelical and least hierarchical of American denominations.

revivalists converted thousands of migrants in the new states of Kentucky, Tennessee, and Ohio. Riding the crest of this wave of religious enthusiasm, Baptist and Methodist preachers traveled around the countryside, organizing hundreds of congregations run by laypeople. Finally, during the 1820s, Lyman Beecher and Charles Grandison Finney led major revivals among Congregationalists and Presbyterians in New England and New York. Religious passions were particularly intense in Rochester, New York, and in other communities along the Erie Canal, giving this region a well-deserved reputation as "the burned-over district."

Religious organizations with female memberships were a key innovation of the so-called Second Great Awakening. For decades, women had formed a majority of the full (or "converted") members of many Congregational churches in New England, and they were numerically powerful in other denominations as well. Encouraged by ministers such as Thomas Bernard of Salem, Massachusetts, these religious women started organizing to achieve specific goals. In Salem, laywomen founded the Female Charitable Society to bring spiritual and material assistance to the needy. In Boston, Mary Webb formed the Female Society for Missionary Purposes. Soon women throughout New England managed local "cent" societies that raised funds for missionary work.

As women took a more public role in religious activities, they encountered increasing hostility from men. "Women have a different *calling*," a pious layman maintained, "they are neither required nor permitted to be exhorters or leaders in public assemblies." The crisis came in 1837, when antislavery advocates Angelina and Sarah Grimké undertook a speaking tour in New England. Ministers and laymen had tolerated, even encouraged, the creation of more than 100 Female Anti-Slavery Societies, but many refused to allow the Grimké sisters — or any women — to lecture to men from a public platform. Rejected as moral equals, many female antislavery activists began a movement for women's rights. "In striving to strike *his* [the black man's] irons off," feminist Abby Kelley recalled, "we found most surely, that *we* were manacled *ourselves*."

The battle over equal rights — for women, artisans, or capitalist entrepreneurs — would be fought on new social terrain. By 1820, the patriarchal family was no longer the building block of the social order, especially in urban areas. As household production declined, the family lost one of its functions, and fathers lost economic leverage over their children. Simultaneously, schools assumed greater responsibility for education. Mechanics' institutes provided vocational training for the children of ambitious artisans and aspiring laborers. Public high schools and private academies educated the offspring of more privileged social groups.

Families lived in a world full of new institutions — schools, hospitals, business corporations, private social clubs, and social reform organizations. If they desired, individuals could orient their lives around these new agencies and enterprises. As a result, the biological and legal ties of family diminished somewhat in importance. Newly married couples were less dependent on their parents for work or social identity, and their children would rely even less on the extended ties of kinship. This new cultural world allowed the emergence of the middle-class family. Its members were more on their own than their ancestors were, linked to the wider society by the bonds of self-interest, voluntary consent, and religious activism. Here, in embryo, was a social environment that would facilitate (even cause) the appearance of modern individualism. As the children of 1820 grew to maturity, they could choose among competing occupations, churches, associations, and values. The human experience encompassed a broader and more diverse social space.

For all of its opportunities, this new middle-class world was hardly a utopia. Children were still born into families that were white or black, poor or rich, farmers or wage-earners. Some were born female, others male. A person's social

An Early Nineteenth-Century Middle-Class Family. This portrait of the Hawley family of Albany, New York, painted in 1801, shows social values in transition. The husband remains dominant (at the center of the composition), and the wife is depicted in a traditional "mothering" pose. However, the children exhibit freedom of movement, suggesting their emergence as individuals. (*Source: Albany Institute of History and Art.*)

origin and gender continued to shape her or his destiny. The daughters of merchants married lawyers; the sons of farmers became laborers, carpenters, bricklayers, or storekeepers. Only the very exceptional or the very fortunate man or woman could rise very far above the social circumstances of birth. In this society, as in any other, a child was not born "free," but came to parents who were members of a specific class, or caste. And parents, for the most part, could do little more than teach their offspring what they themselves had learned, enjoyed, and suffered.

Yet the world of 1820 was far different from that of 1775. To be a white American in 1820 was to be a committed republican, an active Protestant Christian, an advocate of equal rights, and a participant in an increasingly capitalistic economic system. As in all societies, a substantial minority dissented from the dominant system of cultural rules and values. And many more individuals found it difficult to resolve the internal contradictions among these principles. For example, was the entrepreneurially inspired "will" theory of legal contracts compatible with the Christian injunctions to behave fairly and justly in business transactions? Were the economic inequalities inherent in capitalism and the social inequalities of slavery consistent with the republican ideal of equal political rights? These kinds of questions had no easy answers. However, taken together, the trinity of republicanism, Protestantism, and capitalism formed an interlocking web of ideas and institutions that defined the core of American social action. Within these limits, succeeding generations of Americans would struggle to work out their individual lives.

A Final Perspective

From a European point of view, early American society assumed a flat, almost one-dimensional character. Especially before 1740, it lacked a rich, diverse, and complex cultural landscape. Missing were the sharp peaks and subtle contours formed by the slow accretion over long centuries of institutions, architectural styles, and popular traditions. This lack of social complexity was due to the origins of the American settlements as colonial ventures. Some European social classes, political institutions, and religious philosophies did not figure prominently in the migratory process, and others gradually withered in the new American environment. As in all colonies, only a partial and selective transplantation of the old society occurred.

Initially, the lack of historical continuity limited the diversity of the inherited European civilization, simplified its complex rules and customs, reduced its variety and complexity, and narrowed the range of wealth, status, and power. In seventeenth-century America, there was no official aristocracy, no subservient peasantry, no hierarchy of bishops, no craft guilds, and no vital tradition of popular rituals and local festivals.

The colonists by no means lived in a social vacuum, however. From the moment of their arrival, they created political and social institutions that were suited to the new environment. By 1740, a long process of evolutionary change had produced a relatively complex society. This indigenous development took two forms. On the southern mainland (and on the islands of the West Indies), white planters recreated a variant of the traditional class-based society of rural England. They became rich by exploiting the labor of white indentured servants and enslaving tens of thousands of Africans. They became powerful through gaining control of church vestries and political assemblies. To maintain their authority, they married their sons and daughters to social equals, creating a powerful economic and political elite. They used brute force to control their slaves, economic and legal power to intimidate white freeholders and tenants, and political solidarity to contest the authority of royal governors.

In the northern mainland colonies, freeholding farm families dominated rural society, and prosperous merchants controlled the commerce of the seaports. Broad ownership of agricultural property forestalled the appearance of a landed aristocracy, except in the Hudson River Valley of New York. The predominance of freeholding farm families encouraged active participation in local and provincial governments and in strong religious institutions. Northern colonial society was less hierarchical and exploitative than that of the southern colonies. In many respects, it looked more to the American future than to the English past. Its conflicts were not between masters and servants, but between creditors and debtors and among rival ethnic groups.

Beginning in 1740, a series of crises undermined the stability of these established political and social orders. Religious turmoil, war with France, and an economic cycle of boom and bust struck in rapid succession. Britain's sudden imposition of new measures of taxation and control prompted riots, petitions, and the movement for American independence. By 1775, many colonists had repudiated British rule and the traditional monarchical system of government. Many other Americans actively questioned the authority of existing religious institutions and the legitimacy of established political and social distinctions. The struggle for home rule raised the crucial question of who should rule at home.

Between 1776 and 1820, the citizens of the new United States created a republican institutional order. While fighting a financially draining war against Great Britain, they devised effective state constitutions and governments. Subsequently, they organized themselves into a strong national union and began the expansion into the trans-Appalachian west. Americans debated, argued, and even fought bitterly with one another during these years. They were divided into distinct social groups, each seeking to defend or extend its own values and interests.

In the end, the American Revolution had both radical and conservative results. Southern planters retained the ownership of their slaves and a dominant social position in their society, but they had to grant religious liberty to Baptists and greater political authority to white freeholders. Urban artisans and western farmers also exercised greater power in the new republican governments of the north, but merchants and monied landlords remained the most powerful economic groups there.

Subsequent social changes had similarly contradictory results. The first party system was led by gentry and gentlemen, yet the passions of the era mobilized the people at large into quasi-democratic political organizations. Settlement of the western lands simultaneously extended the ideals of a democratic freeholding rural society and the boundaries of the oppressive system of racial slavery. Outwork manufacturing and capitalist-owned factories raised the total wealth of American society, but at the same time made its distribution less equal.

To render any kind of moral judgment on the American historical experience is a delicate and complex task. The men and women who lived between 1600 and 1820 viewed reality from markedly different perspectives. Some inhabitants felt their society constituted, in Tom Paine's phrase, "the last best hope of mankind." But those who bequeathed such optimistic statements to posterity often came from privileged circumstances. As a rule, they were not native American, Afro-American, poor, or female. Yet such unprivileged men and women constituted, at any given time, a majority of all living Americans. The true feelings of these historically disadvantaged groups appear less frequently in the written record. Most of them probably confronted the world with a mixture of hope and bitterness, resigned to the harsh reality of the present while perhaps dreaming of a better future to come.

Our present perspective on early American society must also be ambivalent. It is possible to say, with some confidence, that most Europeans benefited from their transplantation to North America between 1600 and 1820. They enjoyed more favorable material circumstances and wider spiritual opportunities than did the inhabitants of most preindustrial societies. Yet European migrants achieved this success in part by dispossessing native Americans of their lands and subjecting Africans to a system of slavery as damaging and degrading as that found in any so-called civilized society. Even among the favored white population in early America, there was both conflict and consensus, injustice and equity. Some individuals and social groups profited far more than others, often by skill and energy that commanded widespread respect, but sometimes by means that bred extensive resentment.

The legacy of this contradictory past extends even into the political and social struggles of the present. We are all, in some measure, the products of this history, no matter how recent the arrival of our ancestors. And we are also the actors who will help to determine the final outcome and ultimate moral significance of this history. So that, in the end, there is no end at all.

Suggestions for Further Reading

This bibliography is intended as both a guide to important secondary material for the interested student and an acknowledgment of our intellectual debts to those scholars whose works we have used. The arrangement is by chapter (and by sections within each chapter). In many cases, a given article or book is mentioned only once but has provided information and ideas used in a number of different contexts. The abbreviations used for the most frequently cited publications are as follows:

AHR *American Historical Review*
CSSH *Comparative Studies in Society and History*
EHR *Economic History Review,* Second Series
JAH *Journal of American History*
JEBH *Journal of Economic and Business History*
JEH *Journal of Economic History*
JIH *Journal of Interdisciplinary History*
JSH *Journal of Southern History*
JSocH *Journal of Social History*
P&P *Past and Present*
PAH *Perspectives in American History*
WMQ *William and Mary Quarterly,* Third Series

Chapter One

Peter Laslett, *The World We Have Lost* (London: Methuen, 1965), paints a brilliant portrait of English preindustrial society, and Pierre Goubert, *Louis XIV and Twenty Million Frenchmen* (New York: Pantheon Books, 1970), does the same for France. Other important surveys of early modern England include Keith Wrightson, *English Society, 1580–1680* (New Brunswick, N.J.: Rutgers University Press, 1968); Carl Bridenbaugh, *Vexed and Troubled Englishmen, 1590–1642* (New York: Oxford University Press, 1968); Wallace Notestein, *The English People on the Eve of Colonization, 1603–1630* (New York: Harper & Row, 1954); and Christopher Hill, *The Century of Revolution, 1603–1714* (New York: W. W. Norton, 1982).

John E. Neale's *Queen Elizabeth I, A Biography* (New York: Doubleday, 1957) is a readable study; for her successors, see Godfrey Davies, *The Early Stuarts, 1603–1660*

(Oxford: The Clarendon Press, 1959). Charles H. George and Katherine George discuss the background of Puritanism in *The Protestant Mind of the English Reformation, 1570–1640* (Princeton, N.J.: Princeton University Press, 1961), as does Patrick Collinson in *The Elizabethan Puritan Movement* (Berkeley, Calif.: University of California Press, 1967). Michael Walzer, *The Revolution of the Saints: A Study in the Origins of Radical Politics* (New York: Atheneum, 1968), carries the story of Puritanism forward.

Earl J. Hamilton, *American Treasure and the Price Revolution in Spain, 1501–1650* (New York: Octagon Books, reprint, 1965), first traced the impact of American gold and silver; an expanded version of the story appears in T. A. Aston (ed.), *Crisis in Europe, 1560–1660* (Garden City, N.J.: Anchor Books, 1967). Specific studies of social groups include Lawrence Stone, "Social Mobility in England, 1500–1700," *P&P* 33 (1966), and *The Crisis of the Aristocracy, 1558–1641* (Oxford: The Clarendon Press, 1965); Mildred L. Campbell, *The English Yeoman under Elizabeth and the Early Stuarts* (New York: Barnes & Noble, reprint, 1960); W. G. Hoskins, *The Midland Peasant: The Economic and Social History of a Leicestershire Village* (New York: St. Martin's, 1957). See also Peter Clark (ed.), *The Early Modern Town* (New York: Longman, 1976); Margaret Spufford, *Contrasting Communities: English Villagers in the Sixteenth and Seventeenth Centuries* (New York: Cambridge University Press, 1974); and B. H. Slichter Van Bath, *The Agrarian History of Western Europe, A.D. 500–1850* (New York: St. Martin's, 1963).

David B. Quinn, *Raleigh and the British Empire* (Harmondsworth, Eng.: Penguin Books, 1973), and Alfred L. Rowse, *The Expansion of Elizabethan England* (New York: Scribner's, 1972), offer comprehensive accounts of early English colonization.

The settlement of Virginia is described in Alden Vaughan, *American Genesis: Captain John Smith and the Founding of Virginia* (Boston: Little Brown, 1975); Edmund S. Morgan, *American Slavery, American Freedom: The Ordeal of Colonial Virginia* (New York: Norton, 1975); and Darrett and Anita Rutman, "Of Agues and Fevers: Malaria in the Early Chesapeake," *WMQ* 33 (1976). Two striking conceptual frameworks for the study of early Chesapeake society are Sigmund Diamond, "From Organization to Society: Virginia in the Seventeenth Century," in Paul Goodman (ed.), *Essays in American Colonial History* (Freeport, N.Y.: Books for Libraries Press, 1970); and Bernard Bailyn, "Politics and Social Structure in Virginia," in James M. Smith (ed.), *Seventeenth Century America: Essays in Colonial History* (Chapel Hill, N.C.: University of North Carolina Press, 1959). See also Jack Greene, "Foundations of Political Power in the Virginia House of Burgesses, 1720–1776," *WMQ* 16 (1959). For material on Maryland, see Aubrey C. Land, *Colonial Maryland: A History* (Millwood, N.Y.: KTO Press, 1981); Lois G. Carr and Lorena S. Walsh, "The Planter's Wife: The Experience of White Women in Seventeenth-Century Maryland," *WMQ* 34 (1977); and the essays in A. C. Land, L. G. Carr, and E. Papenfuse (eds.), *Law, Society, and Politics in Early Maryland* (Baltimore: Johns Hopkins University Press, 1977).

Edmund S. Morgan's *The Puritan Dilemma: The Story of John Winthrop* (Boston: Little Brown, 1958) is a fine biography, and Alan Simpson's *Puritanism in Old and New England* (Chicago: University of Chicago Press, 1961) provides an overview of this important religious and social movement. David D. Hall (ed.), *The Antinomian Controversy, 1636–1638: A Documentary History* (Middletown, Conn.: Wesleyan University Press, 1968); Emery Battis, *Saints and Sectaries: Anne Hutchinson and the Antinomian Controversy in the Massachusetts Bay Colony* (Chapel Hill, N.C.: University of North Carolina Press, 1962); and Lyle Koehler, *A Search for Power: The "Weaker Sex" in Seventeenth-

Century New England (Urbana: University of Illinois Press, 1980) offer different perspectives on Anne Hutchinson. Ola Winslow, *Master Roger Williams: A Biography* (New York: Macmillan, 1957); Perry Miller, *Roger Williams* (New York: Atheneum, 1962); and Edmund S. Morgan, *Roger Williams: The Church and the State* (New York: Harcourt, Brace & World, 1967), provide contrasting portraits of the Separatist leader.

Concerning the Puritan state and Puritan society, see the essay by Perry Miller in *Errand into the Wilderness* (New York: Harper & Row, 1964); Darrett Rutman, *Winthrop's Boston: Portrait of a Puritan Town, 1630–1649* (Chapel Hill, N.C.: University of North Carolina Press, 1965); and Timothy H. Breen, "Persistent Localism: English Social Change and the Shaping of New England Institutions," *WMQ* 32 (1975). George Langdon, Jr., *Plymouth Colony: A History of New Plymouth, 1620–1691* (New Haven, Conn.: Yale University Press, 1966), and John Demos, *A Little Commonwealth: Family Life in Plymouth Colony* (New York: Oxford University Press, 1970), explore the American adventures of the Pilgrims. Governor William Bradford's *Of Plymouth Plantation, 1620–1647* (Franklin Center, Pa.: Franklin Library, 1983) offers an eyewitness account.

Francis Jennings, *The Invasion of America: Indians, Colonialism and the Cant of Conquest* (Chapel Hill, N.C.: University of North Carolina Press, 1975), provides a stimulating discussion of cultural conflict that takes issue with the pro-European interpretation advanced by Alden Vaughan in *The New England Frontier: Puritans and Indians, 1620–1675* (Boston: Little Brown, 1965). Gary B. Nash, *Red, White, and Black: The Peoples of Early America* (Englewood Cliffs, N.J.: Prentice Hall, 2nd ed., 1982), provides a fine comparative survey of Indian-European interaction. An important recent interpretation is that of William Cronon, *Changes in the Land: Indians, Colonists, and the Ecology of New England* (New York: Hill & Wang, 1983). Other significant studies include James Axtell, *The European and the Indian: Essays in the Ethnohistory of Colonial North America* (New York: Oxford University Press, 1981); Neal Salisbury, *Manitou and Providence: Indians, Europeans, and the Making of New England, 1500–1643* (New York: Oxford University Press, 1982); and Karen Ordahl Kupperman, *Settling with the Indians: The Meeting of English and Indian Cultures in America, 1580–1640* (Totowa, N.J.: Rowman and Littlefield, 1980). James H. Merrell, "The Indians' New World: The Catawba Experience," *WMQ* 41 (1984), offers a fresh perspective on the reactions of native Americans to the European invasion.

Chapter Two

W. A. Speck, "The International and Imperial Context," in Jack P. Greene and J. R. Pole (eds.), *Colonial British America: Essays in the New History of the Early Modern Era* (Baltimore: Johns Hopkins University Press, 1983), offers a helpful overview that has informed this book's discussion of English mercantilism. The intellectual background of mercantilism is analyzed by Joyce O. Appleby in *Economic Thought and Ideology in Seventeenth-Century England* (Princeton, N.J.: Princeton University Press, 1978). For the development of mercantilist policy, see Theodore K. Rabb, *Enterprise and Empire: Merchants and Gentry Investment in the Expansion of England, 1575–1630* (Cambridge, Mass.: Harvard University Press, 1967); J. E. Farnell, "The Navigation Act of 1651, the First Dutch War and the London Merchant Community," *EHR* 32 (1974); and Jack M. Sosin, *English America and the Restoration Monarchy of Charles II: Transatlantic Politics, Commerce, and Kinship* (Lincoln, Neb.: University of Nebraska Press, 1980).

The portrait of New England town life draws on many of the community studies published since the 1960s. Sumner Chilton Powell's *Puritan Village: The Formation of a New England Town* (Middletown, Conn.: Wesleyan University Press, 1963) was not only the first, but also one of the best in showing the cultural continuity from England to New England. David Grayson Allen's *In English Ways: The Movement of Societies and the Transferral of Local Law and Custom to Massachusetts Bay in the Seventeenth Century* (Chapel Hill, N.C.: University of North Carolina Press, 1981) is an excellent study of five New England towns and reveals the diversity created by the transplantation of English village patterns. See also T. H. Breen, "Persistent Localism: English Social Change and the Shaping of New England Institutions," *WMQ* 32 (1975); and John J. Waters, Jr., "Hingham, Massachusetts, 1631–1661: An East Anglian Oligarchy in the New World," *JSocH* 1 (1967–68). The emphasis on the self-conscious insularity of New England towns is strongest in Kenneth A. Lockridge's *A New England Town, The First Hundred Years: Dedham, Massachusetts, 1636–1736* (New York: W. W. Norton, 1970). Michael Zuckerman, "The Social Context of Democracy in Massachusetts," *WMQ* 25 (1968), explores the political implications of the commitment to collective unity.

The daily and seasonal rhythms of village life are revealed by E. P. Thompson in "Time, Work Discipline, and Industrial Capitalism," *P&P* 38 (1967). Valuable descriptions of the routines of New England farm life are found in Percy Wells Bidwell and John I. Falconer, *History of Agriculture in the Northern United States, 1620–1860* (Washington, D.C.: Carnegie Institute, 1925); and Howard S. Russell, *A Long, Deep Furrow: Three Centuries of Farming in New England* (Hanover, N.H.: University Press of New England, 1976). Important to the analysis of growth and change are certain demographic studies of New England towns. Philip Greven offers the most detailed demographic history of a New England town in *Four Generations: Population, Land, and Family in Colonial Andover, Massachusetts* (Ithaca, N.Y.: Cornell University Press, 1970). Also important are John Demos, *A Little Commonwealth: Family Life in Plymouth Colony* (New York: Oxford University Press, 1970) and "Families in Colonial Bristol, Rhode Island: An Exercise in Historical Demography," *WMQ* 25 (1968); Kenneth A. Lockridge, "The Population of Dedham, Massachusetts, 1636–1736," *EHR* 19 (1966); Susan L. Norton, "Population Growth in Colonial America: A Study of Ipswich, Massachusetts," *Population Studies* 25 (1971); and Daniel Scott Smith, "The Demographic History of Colonial New England," *JEH* 32 (1972). The relationship between population growth and church membership is suggested by Darrett B. Rutman in *Winthrop's Boston: Portrait of a Puritan Town, 1630–1649* (Chapel Hill, N.C.: University of North Carolina Press, 1965). Edmund S. Morgan, *Visible Saints: The History of a Puritan Idea* (Ithaca, N.Y.: Cornell University Press, 1965), analyzes the significance of the Halfway Covenant, as does Robert Pope, *The Half-Way Covenant: Church Membership in Puritan New England* (Princeton, N.J.: Princeton University Press, 1969). The implications of Stoddard's open communion are drawn from Paul R. Lucas, *Valley of Discord: Church and Society along the Connecticut River, 1636–1725* (Hanover, N.H.: University Press of New England, 1976); James P. Walsh, "Solomon Stoddard's Open Communion: A Reexamination," *New England Quarterly* 43 (1970); and Patricia Juneau Tracy, *Jonathan Edwards, Pastor: Religion and Society in Eighteenth-Century Northampton* (New York: Hill and Wang, 1979).

The social and economic implications of population growth are derived from Kenneth A. Lockridge's "Land, Population, and the Evolution of New England Society, 1630–1790," *P&P* 39 (1968). Michael Zuckerman, *Peaceable Kingdoms: New England Towns in the Eighteenth Century* (New York: Knopf, 1970), and James A. Henretta, "The

Morphology of New England Society in the Colonial Period," *JIH* 2 (1971), provide the most useful analyses of the cultural conflicts inherent in town divisions. Paul Boyer and Stephen Nissenbaum offer an extreme but intriguing case study in *Salem Possessed: The Social Origins of Witchcraft* (Cambridge, Mass.: Harvard University Press, 1974).

The analysis of social relations in the Chesapeake region, including the discussion of Bacon's Rebellion, generally follows the interpretation offered by Edmund S. Morgan in *American Slavery, American Freedom: The Ordeal of Colonial Virginia* (New York: W. W. Norton, 1975). Also important for the development of this chapter are several of the essays in Thad W. Tate and David L. Ammerman (eds.), *The Chesapeake in the Seventeenth Century: Essays on Anglo-American Society* (Chapel Hill, N.C.: University of North Carolina Press, 1979), especially James Horn, "Servant Emigration to the Chesapeake in the Seventeenth Century"; Carville V. Earle, "Environment, Disease, and Mortality in Early Virginia"; Kevin P. Kelly, "'In dispers'd Country Plantations': Settlement Patterns in Seventeenth-Century Surry County, Virginia"; and Lois Green Carr and Russell R. Menard, "Immigration and Opportunity: The Freedman in Early Colonial Maryland." Winthrop D. Jordan, *White Over Black: American Attitudes Toward the Negro, 1550–1812* (Chapel Hill, N.C.: University of North Carolina Press, 1968), provides a compelling discussion of the legal and cultural background of slavery. Two articles by Russell R. Menard document the increase in the slave population in the late seventeenth century: "The Maryland Slave Population, 1658 to 1730: A Demographic Profile of Blacks in Four Counties," *WMQ* 32 (1975); and "From Servants to Slaves: The Transformation of the Chesapeake Labor System," *Southern Studies* 16 (1977).

In addition to the studies of English imperial policy cited above, several other works discuss the crisis of the late seventeenth century. John Murrin's essay "Political Development," in Greene and Pole, *Colonial British America* [cited above] is a very useful overview. Stephen Saunders Webb, *1676: The End of American Independence* (New York: Knopf, 1984), emphasizes the effect of Bacon's Rebellion on English policy; and David S. Lovejoy, *The Glorious Revolution in America* (New York: Harper & Row, 1972), chronicles the events of the 1680s.

Chapter Three

This chapter owes a great debt to two recent surveys of colonial economic development: Gary M. Walton and James F. Shepherd, *The Economic Rise of Early America* (New York: Cambridge University Press, 1979); and John J. McCusker and Russell R. Menard, *The Economy of British America, 1607–1789* (Chapel Hill, N.C.: University of North Carolina Press, 1985). The McCusker and Menard work is especially valuable; it provides both a useful theoretical framework — the staples approach — and an excellent regional survey of the colonial economy. However, the staples approach does have its limitations, particularly when applied to the eighteenth century. See Jacob M. Price, "The Transatlantic Economy," in Jack P. Greene and J. R. Pole (eds.), *Colonial British America: Essays in the New History of the Early Modern Era* (Baltimore: Johns Hopkins University Press, 1984).

An excellent and extremely readable study of the early West Indian economy is Richard S. Dunn's *Sugar and Slaves: The Rise of the Planter Class in the English West Indies, 1624–1713* (Chapel Hill, N.C.: University of North Carolina Press, 1972). Also important is Richard B. Sheridan's *Sugar and Slavery: An Economic History of the British West Indies, 1623–1775* (Baltimore: Johns Hopkins University Press, 1974). The demography of the West Indies has been explored in brilliant fashion by Philip Curtin in *The Atlantic Slave Trade* (Madison, Wis.: University of Wisconsin Press, 1969) and in "Epidemiology and the Slave Trade," *Political Science Quarterly* 83 (1968). See also two articles by David W. Galenson: "The Slave Trade to the English West Indies, 1673–1724," *EHR* 32 (1979); and "The Atlantic Slave Trade and the Barbados Market, 1673–1723," *JEH* 42 (1982). The nature of West Indian slavery is analyzed by Orlando Patterson in *The Sociology of Slavery* (Cranbury, N.J.: Fairleigh Dickinson University Press, 1970).

Jack P. Greene stresses the connection between the West Indies and the mainland colonies in "Society and Economy in the British Caribbean during the Seventeenth and Eighteenth Centuries," *AHR* 79 (1974). An outstanding study of the early development of the lower mainland is Peter H. Wood's *Black Majority: Negroes in Colonial South Carolina from 1670 through the Stono Rebellion* (New York: W. W. Norton, 1974). See also David L. Coon, "Eliza Lucas Pinckney and the Reintroduction of Indigo Culture in South Carolina," *JSH* 42 (1976). The best survey of the lower South is still that by Verner W. Crane, *The Southern Frontier, 1670–1732* (New York: W. W. Norton, reprint, 1981). Concerning the spread of rice culture to Georgia, see David R. Chesnutt, "South Carolina's Penetration of Georgia in the 1760's: Henry Laurens as a Case Study," *South Carolina Historical Magazine* 73 (1972). The most comprehensive work on the economy of North Carolina is H. Roy Merrens, *Colonial North Carolina in the Eighteenth Century: A Study in Historical Geography* (Chapel Hill, N.C.: University of North Carolina Press, 1964). A. Roger Ekirch, *"Poor Carolina": Politics and Society in Colonial North Carolina, 1729–1776* (Chapel Hill, N.C.: University of North Carolina Press, 1981), offers a brief overview that emphasizes North Carolina's comparative disadvantages with respect to its neighboring colonies.

Studies of the colonial Chesapeake region's tobacco economy are much more numerous. Among important recent books are Edmund S. Morgan, *American Slavery, American Freedom: The Ordeal of Colonial Virginia* (New York: W. W. Norton, 1975); Allan Kulikoff, *Tobacco and Slaves: The Development of Southern Cultures in the Chesapeake, 1600–1800* (Chapel Hill, N.C.: University of North Carolina Press, 1986); Darrett B. Rutman and Anita H. Rutman, *A Place in Time: Middlesex County, Virginia, 1650–1750* (New York: W. W. Norton, 1985); Richard R. Beeman, *The Evolution of the Southern Backcountry: A Case Study of Lunenburg County, Virginia, 1746–1832* (Philadelphia: University of Pennsylvania Press, 1984); and Gloria L. Main, *Tobacco Colony: Life in Early Maryland, 1650–1720* (Princeton, N.J.: Princeton University Press, 1982). David Klingaman, "The Significance of Grain in the Development of the Tobacco Colonies," *JEH* 29 (1969), outlines the gradual transformation in the nature of the planter economy, and Paul G. E. Clemens explores the topic in more detail in *The Atlantic Economy and Colonial Maryland's Eastern Shore: From Tobacco to Grain* (Ithaca, N.Y.: Cornell University Press, 1980). The work of Jacob M. Price analyzes the growth of the international market for tobacco. See especially "The Rise of Glasgow in the Chesapeake Tobacco Trade, 1707–1775," *WMQ* 11 (1954); "The Economic Growth of the Chesapeake and the European Market, 1697–1775," *JEH* 24 (1964); and *Capital and Credit in British Overseas Trade: The View from the Chesapeake* (Cambridge, Mass.: Harvard University Press, 1980). Price also discusses the stultifying effects of tobacco production on Southern urban development in "Economic Function and the Growth of American Port Towns in the Eighteenth Century," *PAH* 8 (1974).

The distinction made between Southern growth and Northern development is derived from Joseph Schumpeter, *The Theory of Economic Development* (trans. by Opie Redvers) (Cambridge, Mass.: Harvard University Press, 1934). The recent literature on Northern commerce begins with Bernard Bailyn's *The New England Merchants in the Seventeenth Century* (Cambridge, Mass.: Harvard University Press, 1955), which is still one of the best studies of mercantile activity on the early colonial period. Stephen C. Innes, *Labor in a New Land: Economy and Society in Seventeenth-Century Springfield* (Princeton, N.J.: Princeton University Press, 1983), provides a valuable study of a prominent merchant family in the western Massachusetts frontier. Peter Dobkin Hall, *The Organization of American Culture, 1700–1900: Private Institutions, Elites, and the Origins of American Nationality* (New York: New York University Press, 1982), extends the analysis of the New England merchant class into the eighteenth century. The effect of economic development on New England port towns is explored by Darrett B. Rutman, *Winthrop's Boston: Portrait of a Puritan Town, 1630–1649* (Chapel Hill, N.C.: University of North Carolina Press, 1965); James A. Henretta, "Economic Development and Social Structure in Colonial Boston," *WMQ* 22 (1965); Richard P. Gildrie, *Salem, Massachusetts, 1626–1683: A Covenant Community* (Charlottesville, Va.: University Press of Virginia, 1975); Paul Boyer and Stephen Nissenbaum, *Salem Possessed: The Social Origins of Witchcraft* (Cambridge, Mass.: Harvard University Press, 1974); and Christine Leigh Heyrman, *Commerce and Culture: The Maritime Communities of Colonial Massachusetts, 1690–1750* (New York: W. W. Norton, 1984). The study that most influenced this chapter's interpretation of the economic role of eighteenth-century Northern cities is Gary B. Nash's *The Urban Crucible: Social Change, Political Consciousness, and the Origins of the American Revolution* (Cambridge, Mass.: Harvard University Press, 1979).

W. A. Speck, "The International and Imperial Context," in Greene and Pole, *Colonial British America* (cited earlier), provides a very helpful introduction to British imperial policy in the eighteenth century. More detailed treatments are available: Thomas Barrow, *Trade and Empire: The British Customs Service in Colonial America, 1660–1775* (Cambridge, Mass.: Harvard University Press, 1967); Ian K. Steele, *Politics and Colonial Policy: The Board of Trade in Colonial Administration, 1696–1720* (Oxford, Eng.: Oxford University Press, 1968); Alison G. Olson, *Anglo-American Politics, 1660–1775: The Relationship Between Parties in England and Colonial America* (Oxford, Eng.: Oxford University Press, 1973); and James A. Henretta, *"Salutary Neglect": Colonial Administration under the Duke of Newcastle* (Princeton, N.J.: Princeton University Press, 1972).

Chapter Four

Edmund S. Morgan's *The Puritan Family: Religion and Domestic Relations in Seventeenth Century New England* (Westport, Conn.: Greenwood Press, reprint, 1980) is an important introduction to the history of the family, although it is based primarily on literary sources. John Demos, *A Little Commonwealth: Family Life in Plymouth Colony* (New York: Oxford University Press, 1970) deals with the topic by adding demographic and neo-Freudian analyses. The importance of education is explored by Bernard Bailyn, *Education in the Forming of American Society* (Chapel Hill, N.C.: University of North Carolina Press, 1960), and Kenneth A. Lockridge, *Literacy in Colonial New England: An Inquiry into the Social Context of Literacy in the Early Modern West* (New York: W. W. Norton, 1974). Philip D. Greven addresses the use of inheritance as a form of family discipline in *Four Generations: Population, Land, and Family in Colonial Andover, Massachusetts* (Ithaca, N.Y.: Cornell University Press, 1970). See also John J. Waters, "Patrimony, Succession, and Social Stability: Guilford, Connecticut, in the Eighteenth Century," *PAH* 10 (1976). An important study of the place of women in colonial New England is that by Laurel T. Ulrich, *Good Wives: Image and Reality in the Lives of Women in Northern New England, 1650–1750* (New York: Oxford University Press, 1980). Also useful is Lyle Koehler's *A Search for Power: The "Weaker Sex" in Seventeenth Century New England* (Champaign, Ill.: University of Illinois Press, 1980). Daniel Scott Smith discusses the significance of women's birth order in "Parental Power and Marriage Patterns: An Analysis of Historical Trends in Hingham, Massachusetts," *Journal of Marriage and the Family* 35 (1973). For studies of the end of marriage, see Alexander Keyssar, "Widowhood in Eighteenth-Century Massachusetts: A Problem in the History of the Family," *PAH* 8 (1974); and Nancy F. Cott, "Divorce and the Changing Status of Women in Eighteenth-Century Massachusetts," *WMQ* 33 (1976).

The literature on Quaker families is more limited, but especially important are the following: Barry J. Levy, "'Tender Plants': Quaker Farmers and Children in the Delaware Valley, 1681–1735," *Journal of Family History* 3 (1978); James T. Lemon, *The Best Poor Man's Country: A Geographical Study of Early Southeastern Pennsylvania* (Baltimore: Johns Hopkins University Press, 1972); James T. Lemon and Gary B. Nash, "The Distribution of Wealth in Eighteenth Century America: A Century of Change in Chester County, Pennsylvania, 1693–1802," *JSocH* 2 (1968); and Stephanie Grauman Wolf, *Urban Village: Population, Community, and Family Structure in Germantown, Pennsylvania, 1683–1800* (Princeton, N.J.: Princeton University Press, 1976).

Two essays by Lois Green Carr offer very useful analyses of the nature of family relations in the early Chesapeake: "The Planter's Wife: The Experience of White Women in Seventeenth-Century Maryland," *WMQ* 34 (1977); and "The Development of the Maryland Orphan's Court, 1654–1715," in Aubrey C. Land, Lois Green Carr, and Edward C. Papenfuse (eds.), *Law, Society, and Politics in Early Maryland* (Baltimore: Johns Hopkins University Press, 1977). Also important is the essay by Darrett B. Rutman and Anita H. Rutman, "Now-Wives and Sons-in-Law: Parental Death in a Seventeenth-Century Virginia County," in Thad W. Tate and David L. Ammerman (eds.), *The Chesapeake in the Seventeenth Century: Essays on Anglo-American Society* (Chapel Hill, N.C.: University of North Carolina Press, 1979). An excellent portrait of the Virginia gentry is Rhys Isaac's *The Transformation of Virginia, 1740–1790* (Chapel Hill, N.C.: University of North Carolina Press, 1982). Gerald W. Mullin, *Flight and Rebellion: Slave Resistance in Eighteenth-Century Virginia* (New York: Oxford University Press, 1972), provides a very good portrait of William Byrd II. For discussions of family life among the Virginia gentry, see Jan Lewis, *The Pursuit of Happiness: Family and Values in Jefferson's Virginia* (New York: Cambridge University Press, 1983); and Daniel Blake Smith, *Inside the Great House: Planter Family Life in Eighteenth-Century Chesapeake Society* (Ithaca, N.Y.: Cornell University Press, 1980).

Many studies of slave life have been done in the past two decades. Among the most useful for the study of the colonial period are Mullin, *Flight and Rebellion*; John W. Blassingame, *The Slave Community: Plantation Life in the Antebellum South* (New York: Oxford University Press, 2nd ed., 1979); Peter Wood, *Black Majority: Negroes in Colonial South Carolina from 1670 through the Stono Rebellion* (New York: W. W. Norton, 1974); Herbert Gutman, *The Black Family in Slavery and Freedom, 1750–1925* (New York: Pantheon, 1976); and Allan Kulikoff, "The Origins of Afro-American Society in Tidewater Maryland and Virginia, 1700 to 1790," *WMQ* 35 (1978). Philip Morgan, "Work and Culture: The Task System and the World of Lowcountry Blacks, 1700 to 1880," *WMQ*

39 (1982), offers an excellent exploration of slaves' ability to determine plantation work patterns.

In addition to the abovementioned *The Transformation of Virginia* by Isaac, important works on the public power of the Southern gentry are A. G. Roeber, "Authority, Law, and Custom: The Rituals of Court Day in Tidewater Virginia, 1720 to 1750," *WMQ* 37 (1980); and Robert M. Weir, "'The Harmony We Were Famous For': An Interpretation of Pre-Revolutionary South Carolina Politics," *WMQ* 26 (1969). Also useful is an unpublished essay by Daniel Scott Smith, "Genealogy, Geography, and the Genesis of Social Structure: Household and Kinship in Early America," which was presented to a conference titled "The Social World of Britain and America, 1600–1820: A Comparison from the Perspective of Social History" held at Williamsburg, Virginia, September 5–7, 1985.

For comparable studies of Northern elites in politics, see Thomas J. Purvis, "'High-Born, Long-Recorded Families': Social Origins of New Jersey Assemblymen," *WMQ* 37 (1980); Patricia U. Bonomi, *A Factious People: Politics and Society in Colonial New York* (New York: Columbia University Press, 1971); Edward M. Cook, Jr., *The Fathers of the Towns: Leadership and Community Structure in Eighteenth-Century New England* (Baltimore: Johns Hopkins University Press, 1976); Robert Zemsky, *Merchants, Farmers, and River Gods: An Essay on Eighteenth-Century American Politics* (Boston: Gambit, 1971); Jere Daniell, "Politics in New Hampshire under Governor Benning Wentworth, 1741–1767," *WMQ* 23 (1966); Mack E. Thompson, "The Ward-Hopkins Controversy and the American Revolution in Rhode Island: An Interpretation," *WMQ* 16 (1959); and Gary B. Nash, *Quakers and Politics: Pennsylvania, 1681–1726* (Princeton, N.J.: Princeton University Press, 1968). Bernard Bailyn, *The Origins of American Politics* (New York: Knopf, 1968), expertly delineates the contours of political life in the eighteenth century, but see also Jack P. Greene, "The Role of the Lower Houses of Assembly in Eighteenth Century Politics," *JSH* 27 (1961) and "Political Mimesis: A Consideration of the Historical and Cultural Roots of Legislative Behavior in the British Colonies in the Eighteenth Century," *AHR* 75 (1969–70). For a more recent synthesis of colonial politics, see John M. Murrin, "Political Development," in Jack P. Greene and J. R. Pole (eds.), *Colonial British America: Essays in the New History of the Early Modern Era* (Baltimore: Johns Hopkins University Press, 1984).

Chapter Five

Richard Hofstadter, *America at 1750* (New York: Knopf, 1971) provides a superb overview of the Great Awakening. For a fine account of the social and religious history of New England, see Richard L. Bushman, *From Puritan to Yankee: Character and the Social Order in Connecticut, 1690–1765* (Cambridge, Mass.: Harvard University Press, 1967). Robert G. Pope, "New England versus the New England Mind: The Myth of Declension," *JSocH* 3 (1969), outlines the changing structure of colonial religion. Two good biographies of revivalists are Patricia Tracy's *Jonathan Edwards, Pastor: Religion and Society in Eighteenth-Century Northampton* (New York: Hill & Wang, 1979), and Christopher Jedrey's *The World of John Cleaveland* (New York: Norton, 1979). An excellent anthology of primary documents is Richard Bushman (ed.), *The Great Awakening: Documents on the Revival of Religion, 1740–1745* (New York: Atheneum, 1970).

Ernest Caufield, "A History of the Terrible Epidemic . . . ," *Yale Journal of Biology and Medicine* 9 (1938–39), suggests the importance of the throat distemper. Gary B. Nash, *The Urban Crucible: Social Change, Political Consciousness, and the Origins of the American Revolution* (Cambridge, Mass.: Harvard University Press, 1979), demonstrates the impact of colonial wars on all aspects of social life. Daniel Scott Smith, "Parental Power and Marriage Patterns: An Analysis of Historical Trends in Hingham, Massachusetts," *Journal of Marriage and the Family* 35 (1973), presents material on premarital pregnancy; to a more limited extent, so does Arthur W. Calhoun in *A Social History of the American Family* (New York: Barnes & Noble, 1945). J. M. Bumsted, "Religion, Finance, and Democracy in Massachusetts: The Town of Norton as a Case Study," *JAH* 57 (1971), relates economic conditions to religious response.

Edwin S. Gaustad, "Society and the Great Awakening in New England," *WMQ* 11 (1954), demonstrates the universal impact of the initial phase of the Awakening, and William G. McLoughlin, *Isaac Backus and the American Pietistic Tradition* (Boston: Little Brown, 1967), traces the subsequent divisions. Rhys Isaac, *The Transformation of Virginia: 1740–1790* (Chapel Hill, N.C.: University of North Carolina Press, 1982), brilliantly captures the spirit of the revival and the social upheaval it produced.

In "Land, Population, and the Evolution of New England Society, 1630–1790," *P&P* 39 (1968), Kenneth Lockridge outlines some of the motive forces behind the movement westward, as does Robert Gross in *The Minutemen and Their World* (New York: Hill & Wang, 1976). John W. Shy, *Toward Lexington: The Role of the British Army in the Coming of the American Revolution* (Princeton, N.J.: Princeton University Press, 1965) deals with the New York tenant revolts, as does Edward Countryman, *A People in Revolution: The American Revolution and Political Society in New York, 1760–1790* (Baltimore: Johns Hopkins University Press, 1981). James A. Henretta, *Salutary Neglect: Colonial Administration Under the Duke of Newcastle* (Princeton, N.J.: Princeton University Press, 1972), explains the new spirit of imperial reform. The concluding chapter of Clarence L. Ver Steeg's *The Formative Years, 1607–1763* (New York: Hill & Wang, 1964) lucidly discusses the Great War for Empire. Fred Anderson, *A People's Army: Massachusetts Soldiers and Society in the Seven Years' War* (Chapel Hill, N.C.: University of North Carolina Press, 1984) offers a new perspective on the war.

Marc Engal, "The Economic Development of the Thirteen Continental Colonies, 1720 to 1775," *WMQ* 32 (1975), provides a good overview, as does Stuart Bruchey, *The Roots of American Economic Growth, 1607–1861: An Essay in Social Causation* (New York: Harper & Row, 1965). See also William S. Sachs, "Agricultural Conditions in the Northern Colonies Before the Revolution," *JEH* 13 (1953). For the relationship between economic development and social attitudes, see Edmund S. Morgan, "The Puritan Ethic and the American Revolution," *WMQ* 24 (1967).

There is no general study of ethnic conflicts in the colonies, but see R. J. Dickson, *Ulster Emigration to Colonial America, 1718–1775* (London: Routledge & Kegan Paul, 1966); Alan Tully, *William Penn's Legacy: Politics and Social Structure in Provincial Pennsylvania, 1726–1755* (Baltimore: Johns Hopkins University Press, 1978); Stephanie G. Wolf, *Urban Village: Population, Community, & Family Structure in Germantown, Pennsylvania, 1683–1800* (Princeton, N.J.: Princeton University Press, 1980); and Richard Alan Ryerson, "Republican Theory and Partisan Reality in Revolutionary Pennsylvania: Toward a New View of the Constitutionalist Party," in Ronald Hoffman and Peter J. Albert (eds.), *Sovereign States in an Age of Uncertainty* (Charlottesville, Va.: University Press of Virginia, 1982). Richard M. Brown, *The South Carolina Regulators* (Cambridge,

Mass.: Harvard University Press, 1963), and R. N. Klein, "Ordering the Backcountry: The South Carolina Regulation," *WMQ* 38 (1981), discuss events in South Carolina, as does Robert M. Weir, "'The Harmony We Were Famous For': An Interpretation of Pre-Revolutionary South Carolina Politics," *WMQ* 26 (1969). For material on North Carolina, see A. Roger Ekirch, "North Carolina Regulators on Liberty and Corruption, 1766–1771," *Perspectives in American History* 11 (1977–78), and Marvin Kay, "The North Carolina Regulation," in Alfred Young (ed.), *The American Revolution: Explorations in the History of American Radicalism* (Dekalb, Ill.: Northern Illinois University Press, 1976).

Chapter Six

Chalmers Johnson, *Revolutionary Change* (Stanford, Calif.: Stanford University Press, 1982), presents a conceptual model of the disintegration of political authority. Charles M. Andrews, *The Colonial Background of the American Revolution* (New Haven, Conn.: Yale University Press, rev. ed., 1931), sets the stage for the revolt, and Edward Countryman, *A People in Revolution: The American Revolution and Political Society in New York, 1760–1790* (Baltimore: Johns Hopkins University Press, 1981), provides a vivid case study of the process of political transformation. Also, Countryman's *The American Revolution* (New York: Hill & Wang, 1985) constitutes a fine synthesis of recent scholarship.

Lawrence Gipson demonstrates the importance of the French and Indian War in *The Coming of the Revolution, 1763–1775* (New York: Harper & Row, 1954), and John Shy, *A People Numerous and Armed: Reflections on the Military Struggle for American Independence* (New York: Oxford University Press, 1976), shows the crucial role played by imperial bureaucrats. Thomas C. Barrow, *Trade and Empire: The British Customs Service in Colonial America: 1660–1775* (Cambridge, Mass.: Harvard University Press, 1967); John L. Bullion, *A Great and Necessary Measure: George Grenville and the Genesis of the Stamp Act, 1763–1765* (Columbia, Mo.: University of Missouri Press, 1983); and Joseph Albert Ernst, *Money and Politics in America, 1755–1775: A Study of the Currency Act of 1764 and the Political Economy of the Revolution* (Chapel Hill, N.C.: University of North Carolina Press, 1973), provide detailed accounts of specific British policies. For broad approaches to the imperial problem, see Ian R. Christie and Benjamin W. Labaree, *Empire and Independence, 1760–1776: A British-American Dialogue on the Coming of the American Revolution* (New York: W. W. Norton, 1976), and Robert W. Tucker and David G. Hendrickson, *The Fall of the First British Empire: Origin of the War of American Independence* (Baltimore: Johns Hopkins University Press, 1982).

Edmund S. Morgan and Helen M. Morgan, *The Stamp Act Crisis: Prologue to Revolution* (Chapel Hill, N.C.: University of North Carolina Press, 1953), provide what is still the best introduction to the events of 1765. Complementary accounts include those by Peter Shaw, *American Patriots and the Rituals of Revolution* (Cambridge, Mass.: Harvard University Press, 1981), and by Pauline Maier, *From Resistance to Revolution: Colonial Radicals and the Development of American Opposition to Britain, 1765–1776* (New York: Random House, 1973). The problem of the colonial mob is treated analytically by Gordon S. Wood, "A Note on Mobs in the American Revolution," *WMQ* 23 (1966), and by Pauline Maier, "Popular Uprisings and Civil Authority in Eighteenth-Century America," *WMQ* 27 (1970). Dirk Hoerder, *Crowd Action in Revolutionary Massachusetts, 1765–1780* (New York: Academic Press, 1977), summarizes the debate and provides additional material, as does Gary B. Nash, *The Urban Crucible: Social Change, Political Consciousness, and the Origins of the American Revolution* (Cambridge, Mass.: Harvard University Press, 1979). A broad general treatment is given by George F. Rudé, *The Crowd in History: A Study of Popular Disturbances in France & England, 1730–1848* (London: Lawrence & Wishart, 1981).

Peter Berger and Thomas Luckmann, *The Social Construction of Reality: A Treatise in the Sociology of Knowledge* (New York: Anchor Books, 1967), demonstrate the complex relationships between ideology and social change. Bernard Bailyn, *The Ideological Origins of the American Revolution* (Cambridge, Mass.: Harvard University Press, 1967), gives a brilliant exposition of the intellectual world of the Patriot leadership. In *The Origins of American Politics* (New York: Knopf, 1968), Bailyn shows how these ideas were imbedded in political institutions. Other discussions of American "republican" ideology include these three: Jack P. Greene, "Political Mimesis: A Consideration of the Historical and Cultural Roots of Legislative Behavior in the British Colonies in the Eighteenth Century," *AHR* 75 (1969–70); J. G. A. Pocock, *The Machiavellian Moment: Florentine Political Thought and the Atlantic Republican Tradition* (Princeton, N. J.: Princeton University Press, 1975); and Caroline Robbins, *The Eighteenth Century Commonwealthman* (Cambridge, Mass.: Harvard University Press, 1959). Two important articles raise objections to an overly intellectual approach: Gordon S. Wood, "Rhetoric and Reality in the American Revolution," *WMQ* 23 (1966); and Marc Engal and Joseph Ernst, "An Economic Interpretation of the American Revolution," *WMQ* 29 (1972).

The roots of religious radical republicanism are outlined by Alan Heimert in *Religion and the American Mind: From the Great Awakening to the Revolution* (Cambridge, Mass.: Harvard University Press, 1966), and by Stephen A. Marini in *Radical Sects of Revolutionary New England* (Cambridge, Mass.: Harvard University Press, 1982). The outlook and motivations of American artisans are described by Eric Foner, *Tom Paine and the American Revolution* (New York: Oxford University Press, 1976); Jesse Lemisch, "Jack Tar in the Streets: Merchant Seamen in the Politics of Revolutionary America," *WMQ* 25 (1968); James Hudson, "An Investigation of the Inarticulate: Philadelphia's White Oaks," *WMQ* 28 (1971); and Bernard Friedman, "The Shaping of Radical Consciousness in Provincial New York," *JAH* 56 (1970).

P. D. G. Thomas, *British Politics and the Stamp Act Crisis: The First Phase of the American Revolution, 1763–1767* (Oxford, Eng.: Clarendon Press, 1975), explains the British response, as do John Brooke, *King George III* (New York: McGraw-Hill, 1972); John Derry, *English Politics and the American Revolution* (London: Dent, 1976); and Colin Bonwick, *English Radicals and the American Revolution* (Chapel Hill, N.C.: University of North Carolina Press, 1977). Four important articles on the economic aspects of the imperial crisis are the following: J. F. Shepherd and G. M. Walton, "Estimates of Invisible Earnings in the Balance of Payments of the British North American Colonies, 1768–1772," *JEH* 29 (1969); Richard B. Sheridan, "The British Credit Crisis of 1772 and the American Colonies," *JEH* 20 (1960); Emory G. Evans, "Planter Indebtedness and the Coming of the Revolution in Virginia," *WMQ* 19 (1962); and Thad W. Tate, "The Coming of the Revolution in Virginia: Britain's Challenge to Virginia's Ruling Class," *WMQ* 19 (1962).

The final stage of the imperial crisis is discussed by Benjamin Labaree, *The Boston Tea Party* (Boston: Northeastern University Press, 1981); Bernard Donoughue, *British Politics and the American Revolution: The Path to War, 1773–1775* (New York: St. Martin's, 1964); and David Ammerman, *In the Common Cause: American Response to the Coercive Acts of 1774* (Charlottesville, Va.: University Press of Virginia, 1974). For accounts of the crucial surge of Patriot sentiment in the countryside, see Richard L. Bushman, "Massachusetts Farmers and the Revolution," in Richard M. Jellison (ed.), *Society, Freedom, and Conscience: The American Revolution in Virginia, Massachusetts and New York* (New York: W. W. Norton, 1976); and Gregory H. Nobles, *Divisions Throughout the Whole: Politics and Society in Hampshire County, Massachusetts, 1740–1775* (New York: Cambridge University Press, 1983).

Chapter Seven

There are many military histories of the American Revolution. Especially useful general works are Don Higginbotham's *The War of American Independence: Military Attitudes, Policies, and Practice, 1763–1789* (New York: Macmillan, 1971) and Piers Mackesy's *The War for America, 1775–1783* (London: Longmans, 1964). Two works by Howard Peckham, *The War for Independence: A Military History* (Chicago: University of Chicago Press, 1958) and *The Toll of Independence: Engagements and Battle Casualties of the Revolution* (Chicago: University of Chicago Press, 1974), also provide valuable information on the military conflict.

The interpretation offered in this chapter has been greatly influenced by works that place the military experience in its broader social context. The best recent overviews are as follows: Edward Countryman, *The American Revolution* (New York: Hill & Wang, 1985); James Kirby Martin, *In the Course of Human Events: An Interpretive Exploration of the American Revolution* (Arlington Heights, Ill.: AHM Publishing, 1979); and James Kirby Martin and Mark E. Lender, *A Respectable Army: The Military Origins of the Republic, 1763–1789* (Arlington Heights, Ill.: Harlan Davidson, 1982). In two brief essays, "A New Look at the Colonial Militia" and "Hearts and Minds in the American Revolution: The Case of 'Long Bill' Scott and Peterborough, New Hampshire," both in John Shy, *A People Numerous and Armed* (New York: Oxford University Press, 1976), John Shy explores the situation of the common soldier in the colonial era and the Revolution. A fine analysis of the social and cultural implications of the military experience is Charles Royster's *A Revolutionary People at War: The Continental Army and American Character* (Chapel Hill, N.C.: University of North Carolina Press, 1979). See also E. Wayne Carp, *To Starve the Army at Pleasure: Continental Army Administration and American Political Culture, 1775–1783* (Chapel Hill, N.C.: University of North Carolina Press, 1984).

For the ideological background of the independence movement, see Bernard Bailyn's *The Ideological Origins of the American Revolution* (Cambridge, Mass.: Harvard University Press, 1967). Eric Foner provides the best treatment of Tom Paine's politics in *Tom Paine and Revolutionary America* (New York: Oxford University Press, 1976) and "Tom Paine's Republic: Radical Ideology and Social Change," in Alfred F. Young (ed.), *The American Revolution: Explorations in the History of American Radicalism* (DeKalb, Ill.: Northern Illinois University Press, 1976). Jack N. Rakove traces the development of the Declaration of Independence in the Continental Congress in "The Decision for American Independence: A Reconstruction," *PAH* 10 (1976). Carl Becker's *The Declaration of Independence: A Study in the History of Political Ideas* (New York: Knopf, 1922) is still an important analysis of the ideology of the Declaration. Now, however, it should be read in conjunction with Garry Wills's *Inventing America: Jefferson's Declaration of Independence* (New York: Doubleday, 1978), which downplays the Lockean background and stresses the significance of Scottish Enlightenment thinkers.

A good general treatment of Loyalists is William H. Nelson's *The American Tory* (Westport, Conn.: Greenwood Press, reprint, 1980). See also Wallace Brown, *The Good Americans: The Loyalists in the American Revolution* (New York: Morrow, 1969), and Robert McCluer Calhoon, *The Loyalists in Revolutionary America, 1760–1781* (New York: Harcourt Brace Jovanovich, 1973). The British attempt to recruit blacks for the Loyalist side is discussed by Ira Berlin, "The Revolution in Black Life," in Young (ed.), *The American Revolution.* Two essays by John Shy, "Armed Loyalism: The Case of the Lower Hudson Valley" and "The Military Conflict Considered as a Revolutionary War," both in *A People Numerous and Armed,* explore the role of Loyalists in the armed conflict. Shy's work emphasizes the extent to which the outbreak of warfare affected the balance of political allegiances in a particular area. Joseph S. Tiedemann follows this line of analysis in "Patriots by Default: Queens County, New York, and the British Army, 1776–1783," *WMQ* 43 (1986). For material on Loyalism in New York, see Staughton Lynd, "Who Should Rule at Home? Dutchess County, New York, in the American Revolution," *WMQ* 18 (1961), which should be read in conjunction with Sung Bok Kim, "Impact of Class Relations and Warfare in the American Revolution: The New York Experience," *JAH* 69 (1982).

The initial impact of the revolution on the economic and political life of other Northern communities is discussed in several works: Richard Buel, Jr., *Dear Liberty: Connecticut's Mobilization for the Revolutionary War* (Middletown, Conn.: Wesleyan University Press, 1980); Gregory H. Nobles, *Divisions Throughout the Whole: Politics and Society in Hampshire County, Massachusetts, 1740–1775* (New York: Cambridge University Press, 1983); Christopher Jedrey, *The World of John Cleaveland: Family and Community in Eighteenth-Century New England* (New York: W. W. Norton, 1979); and Robert A. Gross, *The Minutemen and Their World* (New York: Hill & Wang, 1976). Gross's insights about the economy of rural communities provide a useful complement to the discussion by Oscar Handlin and Mary Handlin, "Revolutionary Economic Policy in New England," *WMQ* 4 (1947). See also Anne Bezanson, "Inflation and Controls, Pennsylvania, 1774–1779," *JEH* (supp.) 8 (1948), and Elizabeth Cometti, "Inflation in Revolutionary Maryland," *WMQ* 8 (1951). For a valuable survey of Northern cities in the early part of the Revolution, see Gary B. Nash, *The Urban Crucible: Social Change, Political Consciousness, and the Origins of the American Revolution* (Cambridge, Mass.: Harvard University Press, 1979). Three other studies deal with urban economies in the Revolution, especially the situation of working people: Charles Olton, *Artisans for Independence: Philadelphia Mechanics and the American Revolution* (Syracuse, N.Y.: Syracuse University Press, 1975); Billy G. Smith, "The Material Lives of Laboring Philadelphians, 1750–1800," *WMQ* 38 (1981); and Dirk Hoerder, *Crowd Action in Revolutionary Massachusetts, 1765–1780* (New York: Academic Press, 1977). Hoerder's emphasis on popular regulation of prices builds on E. P. Thompson's important essay, "The Moral Economy of the English Crowd in the Eighteenth Century," *P&P* 50 (1971), and on George Rudé's *The Crowd in History* (New York: John Wiley and Sons, 1964).

The discussion of war in the South draws heavily on Ronald Hoffman, "The Disaffected in the Revolutionary South," in Young (ed.), *The American Revolution.* See also Ronald Hoffman, *A Spirit of Dissension: Economics, Politics, and the Revolution in Maryland* (Baltimore: Johns Hopkins University Press, 1974); Richard R. Beeman, *The Evolution of the Southern Backcountry: A Case Study of Lunenburg County, Virginia, 1746–1832* (Philadelphia: University of Pennsylvania Press, 1984); and Ronald Hoffman, Thad W. Tate, and Peter J. Albert (eds.), *An Uncivil War: The Southern Backcountry during the American Revolution* (Charlottesville, Va.: University Press of Virginia, 1985).

Chapter Eight

The nature of political change in the states is explored in meticulous — and masterful — detail in Gordon Wood's *The Creation of the American Republic, 1776–1787* (Chapel Hill, N.C.: University of North Carolina Press, 1969). Jackson Turner Main also provides a valuable survey in *The Sovereign States, 1775–1783* (New York: New Viewpoints, 1973) and, more briefly, in "Government by the People: The American Revolution and the Democratization of the Legislatures," *WMQ* 23 (1966). See also Willi Paul Adams, *The First American Constitutions: Republican Ideology and the Making of the State Constitutions in the Revolutionary Era* (Chapel Hill, N.C.: University of North Carolina Press, 1980). The case of Massachusetts is covered thoroughly by Ronald M. Peters, Jr., in *The Massachusetts Constitution of 1780: A Social Compact* (Amherst, Mass.: University of Massachusetts Press, 1978).

The discussion of the status of women was influenced by Joan Hoff Wilson, "The Illusion of Change: Women and the American Revolution," in Alfred F. Young (ed.), *The American Revolution: Explorations in the History of American Radicalism* (Dekalb, Ill.: Northern Illinois University Press, 1976). The most valuable studies of women in the Revolutionary era are the following: Linda K. Kerber, *Women of the Republic: Intellect and Ideology in Revolutionary America* (Chapel Hill, N.C.: University of North Carolina Press, 1980); Nancy F. Cott, *The Bonds of Womanhood: "Woman's Sphere" in New England, 1780–1835* (New Haven, Conn.: Yale University Press, 1977); and Mary Beth Norton, *Liberty's Daughters: The Revolutionary Experience of American Women, 1750–1800* (Boston: Little Brown, 1980). Norton gives a suggestive analysis of the significance of fertility decline on women's lives, as does Robert V. Wells in "Family History and Demographic Transition," *JSocH* 9 (1975).

Ira Berlin offers a brief but solid overview of the situation for blacks in "The Revolution in Black Life," in Young's *The American Revolution*. Benjamin Quarles's *The Negro in the American Revolution* (Chapel Hill, N.C.: University of North Carolina Press, 1961) is also important. Concerning the issue of antislavery thought in the Revolutionary era, see Winthrop D. Jordon, *White over Black: American Attitudes toward the Negro, 1550–1812* (Chapel Hill, N.C.: University of North Carolina Press, 1968); David Brion Davis, *The Problem of Slavery in the Age of Revolution, 1770–1823* (Ithaca, N.Y.: Cornell University Press, 1975); and William W. Freehling, "The Founding Fathers and Slavery," *AHR* 77 (1972). An excellent discussion of Gabriel's Rebellion is found in Gerald W. Mullin's *Flight and Rebellion: Slave Resistance in Eighteenth-Century Virginia* (New York: Oxford University Press, 1972).

Curtis P. Nettels, *The Emergence of a National Economy, 1775–1815* (New York: Holt, Rinehart and Winston, 1962), provides a good survey of economic conditions in the Revolutionary era. A shorter but very useful survey is that by John J. McCusker and Russell R. Menard, *The Economy of British America, 1607–1789* (Chapel Hill, N.C.: University of North Carolina Press, 1985). David P. Szatmary, *Shays' Rebellion: The Making of an Agrarian Insurrection* (Amherst, Mass.: University of Massachusetts Press, 1980), explains the "chain of debt" in post-war New England and also provides an excellent analytical narrative of the insurrection. Robert Morris's critique of the economy was taken from James Kirby Martin, *In the Course of Human Events* (Arlington Heights, Ill.: AHM Publishing, 1979).

James Kirby Martin also provides a worthwhile examination of the Constitution-making process, but the most detailed narrative is still Max Farrand's *The Framing of the Constitution of the United States* (New Haven, Conn.: Yale University Press, 1913). Cecilia M. Kenyon's "Men of Little Faith: The Anti-Federalists on the Nature of Representative Government," *WMQ* 12 (1955) is likewise a valuable older work. For a fuller treatment of the Antifederalists, see Jackson Turner Main, *The Antifederalists: Critics of the Constitution, 1781–1788* (Chapel Hill, N.C.: University of North Carolina Press, 1961). Main's *Political Parties before the Constitution* (Chapel Hill, N.C.: University of North Carolina Press, 1973) discusses the ratification controversy.

Chapter Nine

E. James Ferguson, *The Power of the Purse: A History of American Public Finance, 1776–1790* (Chapel Hill, N.C.: University of North Carolina Press, 1961), and Dale W. Forsythe, *Taxation and Political Change in the Young Nation, 1781–1833* (New York: Columbia University Press, 1977), trace the evolution of the American fiscal system. Broadus Mitchell, *Alexander Hamilton: A Concise Biography* (New York: Oxford University Press, 1976), provides a detailed analysis of Hamilton's policies.

Seymour Martin Lipset, *The First New Nation: The United States in Historical and Comparative Perspective* (New York: W. W. Norton, 1979), takes a stimulating approach to the political structure of the early republic. More detailed studies of the first party system include the following: Joseph Charles, *The Origins of the American Party System: Three Essays* (Chapel Hill, N.C.: University of North Carolina Press, 1956); Noble E. Cunningham, Jr., *The Jeffersonian Republicans: The Formation of Party Organization, 1789–1801* (Chapel Hill, N.C.: University of North Carolina Press, 1957); and William N. Chambers, *Political Parties in a New Nation: The American Experience, 1776–1809* (New York: Oxford University Press, 1963). Richard Hofstadter, *The Idea of a Party System: The Rise of Legitimate Opposition in the United States, 1780–1840* (Berkeley, Calif.: University of California Press, 1969), offers a convincing overview. Alfred F. Young, *The Democratic Republicans of New York: The Origins, 1763–1797* (Chapel Hill, N.C.: University of North Carolina Press, 1967), provides a fine study of one state.

Useful theoretical approaches to the structure of politics are as follows: Francis X. Sutton, "Representation and the Nature of Political Systems," *CSSH* 2 (1959); Alex Weingrod, "Patrons, Patronage, and Political Parties," *CSSH* 10 (1968); and Paul Goodman, "The First American Party System," in William N. Chambers and Walter D. Burnham (eds.), *The American Party System: Stages of Political Development* (New York: Oxford University Press, 1975). See also Ronald P. Formisano, *The Transformation of Political Culture: Massachusetts Parties, 1790's–1840's* (New York: Oxford University Press, 1983).

Stephen Kurtz, *The Presidency of John Adams* (Philadelphia: University of Pennsylvania Press, 1957), deals with the political struggles of the late 1790s, as do Leonard W. Levy, *Legacy of Suppression: Freedom of Speech and Press in Early American History* (Cambridge, Mass.: Harvard University Press, 1960), and James M. Smith, *Freedom's Fetters: Sedition Laws and American Civil Liberties* (Ithaca, N.Y.: Cornell University Press, 1966).

Dumas Malone, *Jefferson the President* (Boston; Little Brown, 2 vols., 1970, 1974), gives a masterful treatment to political events. Adrienne Koch, *The Philosophy of Thomas Jefferson* (Chicago: Quadrangle Books, 1964); Drew McCoy, *The Elusive Republic: Political Economy in Jeffersonian America* (Chapel Hill, N.C.: University of North Carolina Press, 1980); and Lance Banning, *The Jeffersonian Persuasion: Evolution of a Party Ideology* (Ithaca, N.Y.: Cornell University Press, 1980), offer treatments of Jefferson's political outlook.

Oscar and Mary Handlin's *Commonwealth: A Study of the Role of Government in the American Economy, Massachusetts, 1774–1861* (Cambridge, Mass.: Harvard University Press, rev. ed., 1969) is still the best study of state mercantilism. Other accounts include Louis Hartz, *Economic Policy and Democratic Thought: Pennsylvania, 1776–1860* (Gloucester, Mass.: Peter Smith, reprint, n.d.); Milton S. Heath, *Constructive Liberalism: The Role of the State in Economic Development in Georgia to 1860* (Cambridge, Mass.: Harvard University Press, 1954); and E. M. Dodd, *American Business Corporations Until 1860 with Special Reference to Massachusetts* (Cambridge, Mass.: Harvard University Press, 1954). For a survey of the literature, see Harry N. Scheiber, "Government and the Economy: Studies of the 'Commonwealth' Policy in Nineteenth-Century America," *JIH* 3 (1972). Frank W. Taussig, *The Tariff History of the United States* (East Orange, N.J.: Kelley, 8th ed., 1931), traces the rise of protectionist policies. For Hamilton's view, see John R. Nelson, Jr., "Alexander Hamilton and American Manufacturing: A Re-examination," *JAH* 65 (1979). For material on state investments in canals, see H. Jerome Cranmer, "Canal Investment, 1815–1860," in National Bureau of Economic Research, *Trends in the American Economy in the Nineteenth Century* (Princeton, N.J.: Princeton University Press, 1960); and Harry N. Scheiber, *Ohio Canal Era: A Case Study of Government and the Economy, 1820–1861* (Athens, Ohio: Ohio University Press, 1969).

Bray Hammond, *Banks and Politics in America from the Revolution to the Civil War* (Princeton, N.J.: Princeton University Press, 1957), outlines the development of the banking system. For the activities of a typical merchant capitalist, see Stuart Bruchey, *Robert Oliver: Merchant of Baltimore, 1783–1819* (Baltimore: Johns Hopkins University Press, 1956). The emergence of specialized financial institutions is covered by H. M. Larson, "S. and M. Allen — Lottery, Exchange, and Stock Brokerage," *JEBH* 3 (1931), and by Arthur H. Cole, "The Evolution of the Foreign Exchange Market of the United States," *JEBH* 1 (1929). Douglass C. North, "International Capital Flows and the Development of the American West," *JEH* 16 (1956), demonstrates the importance of foreign investment in American development, and the results are indicated by Paul A. David, "The Growth of Real Product in the United States Before 1840: New Evidence, Controlled Conjectures," *JEH* 27 (1967), and Robert E. Gallman, "The Pace and Pattern of American Economic Growth," in Lance E. Davis (ed.), *American Economic Growth: An Economist's History of the United States* (New York: Harper & Row, 1972). For discussions of the growing social inequality, see Alice Hanson Jones, *Wealth of a Nation To Be: The American Colonies on the Eve of the Revolution* (New York: Columbia University Press, 1980), and Peter H. Lindert and Jeffrey Williamson, *Long-Term Trends in American Wealth Inequality* (Madison, Wis.: University of Wisconsin Press, 1977).

Lawrence M. Friedman's *A History of American Law* (New York: Simon & Schuster, 1973) is a general survey, and James Williard Hurst's *Law and Social Process in United States History* (New York: Da Capo Press, 1972) is a stimulating overview. Other important works are as follows: Charles M. Haar (ed.), *The Golden Age of American Law* (New York: George Braziller, 1965); Maxwell Bloomfield, *American Lawyers in a Chang-*

ing Society: 1776–1876 (Cambridge, Mass.: Harvard University Press, 1976); Gerald Gawalt, *The Promise of Power: The Emergence of the Legal Profession in Massachusetts, 1760–1840* (Westport, Conn.: Greenwood Press, 1979); and Daniel Calhoun, *Professional Lives in America: Structure and Aspiration, 1750–1850* (Cambridge, Mass.: Harvard University Press, 1965). Two stimulating interpretations of the changes in legal doctrine and practice that occurred during this period are those of Morton Horwitz, *The Transformation of American Law, 1780–1860* (Cambridge, Mass.: Harvard University Press, 1976), and William E. Nelson, *The Americanization of the Common Law: The Impact of Legal Change on Massachusetts Society, 1760–1830* (Cambridge, Mass.: Harvard University Press, 1976).

The development of constitutional doctrine at the state level can be traced in G. Edward White's *The American Judicial Tradition* (New York: Oxford University Press, 1976) and Leonard Levy's *The Law of the Commonwealth and Chief Justice Shaw* (Cambridge, Mass.: Harvard University Press, 1967). Important works on the United States Supreme Court include the following: R. Kent Newmeyer, *The Supreme Court Under Marshall and Taney* (Arlington Heights, Ill.: Harlan Davidson, 1969); and Francis N. Stites, *Private Interest and Public Gain: The Dartmouth College Case, 1819* (Amherst, Mass.: University of Massachusetts Press, 1972).

Chapter Ten

Curtis R. Nettels, *The Emergence of a National Economy, 1775–1815* (New York: M. E. Sharpe, 1966), provides a comprehensive survey, and Douglass C. North, *The Economic Growth of the United States, 1790–1860* (Englewood Cliffs, N.J.: Prentice-Hall, 1961), stresses the importance of the export economy. Phyllis M. Deane, *The First Industrial Revolution* (New York: Cambridge University Press, 1966), and Donald R. Adams, Jr., "Some Evidence on English and American Wage Rates, 1790–1830," *JEH* 28 (1968) allow a trans-Atlantic comparison. L. M. Sears, *Jefferson and the Embargo* (New York: Octagon Books, 1967), and Bradford Perkins, *Prologue to War: England and the United States, 1805–1812* (Berkeley, Calif.: University of California Press, 1961), discuss the politics of wartime commerce.

James A. Henretta, "The War for Independence and American Economic Development," in Ronald Hoffman et al. (eds.), *The American Economy in the Revolutionary Era* (Charlottesville, Va.: University Press of Virginia, 1987), offers an interpretative survey of internal growth. Diane Lindstrom, *The Economic Development of the Philadelphia Region, 1810–1850* (New York: Columbia University Press, 1978), provides a pathbreaking account that both supplements and revises the important study of George Rogers Taylor, *The Transportation Revolution, 1815–1860* (New York: Harper & Row, 1968). For material on the transition from household to factory production, see the classic study by Rolla M. Tryon, *Household Manufactures in the United States: 1640–1860* (New York: A. M. Kelley, reprint, 1966), as well as recent monographs by Alan Dawley, *Class and Community: The Industrial Revolution in Lynn* (Cambridge, Mass.: Harvard University Press, 1976), and Jonathan Prude, *The Coming of Industrial Order: Town and Factory Life in Rural Massachusetts, 1810–1860* (New York: Cambridge University Press, 1983). An important theoretical statement is made by Michael Merrill in "Cash Is Good to Eat: Self-Sufficiency and Exchange in the Rural Economy of the United States," *Radical History Review* 4 (1977),

For accounts of the lives and culture of workers and artisans, see David Montgomery, "The Working Classes of the Pre-Industrial American City, 1780–1830," *Labor History* 9 (1968); Charles S. Olton, *Artisans for Independence: Philadelphia Mechanics and the American Revolution* (Syracuse, N.Y.: Syracuse University Press, 1975); Billy G. Smith, "The Material Lives of Laboring Philadelphians, 1750 to 1800," *WMQ* 38 (1981); and Sean Wilentz, *Chants Democratic: New York City and the Rise of the American Working Class, 1788–1850* (New York: Oxford University Press, 1984). A fine conceptual view of the coercive character of the market economy is found in Karl DeSchweinitz's "Economic Growth, Coercion, and Freedom," *World Politics* 9 (1957).

The impact of the Revolutionary War on native Americans is considered in general terms by Francis Jennings, "The Indians' Revolution," in Alfred F. Young (ed.), *The American Revolution: Explorations in the History of American Radicalism* (DeKalb, Ill.: Northern Illinois University Press, 1976), and in greater detail by Barbara Graymont in *The Iroquois in the American Revolution* (Syracuse, N.Y.: Syracuse University Press, 1972) and by James H. O'Donnell in *Southern Indians in the American Revolution* (Knoxville, Tenn.: University of Tennessee Press, 1973). Anthony F. C. Wallace, *The Death and Rebirth of the Seneca* (New York: Random House, 1970), brilliantly analyzes the military decline and religious revival of the Iroquois. Francis P. Prucha, *American Indian Policy in the Formative Years* (Lincoln, Neb.: University of Nebraska Press, 1970), surveys federal policy, and Bernard Sheehan, *Seeds of Extinction: Jeffersonian Philanthropy and the American Indian* (New York: W. W. Norton, 1974), explores changes in white attitudes toward native Americans.

Material on regional identity in the Old South is abundant; see Clement Eaton, *The Growth of Southern Civilization* (New York: Harper & Row, 1961), and William R. Taylor, *Cavalier and Yankee: The Old South and the American National Character* (Cambridge, Mass.: Harvard University Press, 1961). Intellectual culture in New England is covered by Van Wyck Brooks, *The Flowering of New England* (Franklin Park, Pa.: Franklin Library, reprint, 1979); Stanley R. Schultz, *The Culture Factory: Boston Public Schools, 1789–1860* (New York: Oxford University Press, 1973); and Peter Dobkin Hall, *The Organization of American Culture, 1700–1900* (New York: New York University Press, 1982). Frederick Jackson Turner, *The Frontier in American History* (Franklin Park, Pa.: Franklin Press, reprint, 1977), posits the existence of a distinct Western identity, and Malcolm J. Rohrbough, *The Trans-Appalachian Frontier: People, Societies, and Institutions, 1775–1850* (New York: Oxford University Press, 1978), suggests the complex forms taken by a migratory culture. Paul Johnson, *A Shopkeepers' Millennium: Society and Revivals in Rochester* (New York: Hill & Wang, 1970), demonstrates the transplantation of New England values to upstate New York communities. John Blassingame, *The Slave Community: Plantation Life in the Antebellum South,* 2nd ed. (New York: Oxford University Press, 1979), suggests the impact of the movement westward and the internal slave trade on Afro-American society.

Important conceptual approaches to the formation of social identity include the following: Godfrey and Monica Wilson, *The Analysis of Social Change* (New York: Cambridge University Press, 1968); Robert Redfield, *The Little Community: Peasant Society and Culture* (Chicago: University of Chicago Press, 1955); Thomas Bender, *Community and Social Change in America* (Baltimore: Johns Hopkins University Press, 1982); and Darrett Rutman, "The Social Web: A Prospectus for the Study of the Early American Community," in William L. O'Neill (ed.), *Insights and Parallels: Problems and Issues of American Social History* (Minneapolis: Burgess Publishing, 1973). Robert A. Nisbet, *The*

Quest for Community (Magnolia, Mass.: Peter Smith, 1983), and Elmon R. Service, "Kinship Terminology and Evolution," *American Anthropologist* 64 (1962), provided other helpful background for this topic.

Important studies of religion and society are as follows: Perry Miller, *The Life of the Mind in America: From the Revolution to the Civil War* (San Diego: Hartcourt, Brace, & World, 1970); William G. McLoughlin, *New England Dissent, 1630–1833: The Baptists and the Separation of Church and State* (Cambridge, Mass.: Harvard University Press, 1971); Philip Greven, *The Protestant Temperament: Patterns of Child-Rearing, Religious Experience, and the Self in Early America* (New York: New American Library, 1979); Mary P. Ryan, *Cradle of the Middle Class: The Family in Oneida County, New York, 1790–1865* (New York: Cambridge University Press, 1981); and Whitney R. Cross, *The Burned-over District: The Social and Intellectual History of Enthusiastic Religion in Western New York, 1800–1850* (New York: Octagon Books, 1981).

INDEX

1 2 3 4 5 6 7 8 9 0